ISO 22000: 2018

FOOD SAFETY MANAGEMENT SYSTEM

GENERIC MODEL

VINDIKA
LOKUNARANGODAGE

2018

ISBN 978-955-35840-0-7

A Generic Model of an ISO 22000:2018 Food Safety Management System developed for the application in the tea industry.

Acknowledgements

At the outset of the ISO 22000 Generic Model, the author gratefully acknowledges the dedicated support of tea manufacturers in Sri Lanka, especially low grown orthodox black tea manufacturers who allowed the use of their facilities for research proposes while evaluating the existing conditions and available systems as well as providing insights of the problem they encounter and their expectations.

It was indeed a pleasure working with Wilehena Tea Factory of Pathma Group Tea Factories, who allowed us to use their facilities for the testing and implementation of earlier developments of the generic model and testing and debugging of the errors.

The author intended to acknowledge the Control Union and Bureau Veritas for their continuous support during the early development process of the generic model auditing, commenting, and suggestions for amendments.

Author highly indebted to Prof. K. K. D. S Ranaweera, Dr. I. Wickramasinghe for their tireless support, motivation, comments, and evaluation of the manuscript.

It was my pleasure to Acknowledge Mr. Alfred F. Chambers of Monachus Consulting, Mr. Anil Kahadwuaarachige at SLSI who's passive contributions were a great support to complete this work earlier than expected.

The author is also deeply indebted to Mr. M.R.M Rafi at Pathma Group, Mr. Manoj Wijekoon, Mr. M.M. Sumudu Priyanjith at Whilehena Tea Factory, without their support, the research works would have been an impossible task. Nonetheless, the author acknowledges Mr. Jinadasa at Andaradeniya Group, Mr. Senevirathne at Mahaliyadda Tea Factory, Mr. Hasitha at Thalgasyaya Tea Factory, and Mr. Shantha at Panilkanda Tea Factory for their support in conducting different research activities in their manufacturing facilities in addition to the participation of gap analysis and internal audits.

The list of tea factories was very long, but without their support, this work would have been not possible where author's sincere thanks go to all personnel, including tea factory owners, tea factory officers, assistant factory officers, various kinds of operators participating in interviews, discussions, and evaluations which further includes smallholding farmers, collectors, managers of companies and factories, auctioneers and auditors as well as quality officers.

Finally, the author acknowledges friends, family, and society, including technology that helps to convert these gross ideas into a manuscript.

Preface

This book is intended to help quality assurance personnel working in the tea industry who struggle to develop their own food safety management systems. The tea industry is one of the oldest as to history, and it is the most consumed beverage after the water. Many in the industry do not intend to develop a better quality assurance system; instead, they are interested in obtaining a certificate for marketing purposes. Most of the factory owners, management staff, and even the consultants argue that there is no much financial gain over achieving a good food safety management system due to the high cost involved while no additional price gains in the auctions. Hence, the generic model was specifically designed to use in orthodox black tea manufacturing facilities. The objective was to increase the proper documentation systems in the tea industry and boost their food safety practices. However, the system can be replicated into other food processing facilities with customization of hazard analysis related documents and adjusting procedures according to the requirements of the production, which must be carried out by a person who is knowledgeable on the subject.

However, current consumer trends are against them, and there are growing concerns over the tea industry due to different food safety incidents. The majority of tea importers are in the developed world, but manufacturing is taking place in the underdeveloped third world where most of the lower educated or under-educated marginalized people are involved in the production of tea. Thus, the generic model intends to help the tea industry through the transformation of existing ISO 22000:2005 FSMS with ISO 22000:2018 or the establishment of brand new systems with appropriate system design, upgraded knowledge at an affordable cost.

Actually, it is not a book but a complete food safety management system that can be implemented as an externally developed combination of control measures and procedures quickly and accurately. Free training programs, additional documentation, and online support with word.docx copies will be provided on demand through virtual connectivity.

Vindika Lokunarangodage
July, 2018.

Table of Content

INTRODUCTION

Management of food safety has become one of the major consumer concerns in the global food market due to various risks associated with different manufacturing processes as well as the rapid expansion of ready-to-eat foods. A food safety management is a set of interrelated or interacting elements used for the establishment of a food safety policy and objectives that can deliver consumer perceptions through controlling the activities of a food manufacturing process. The ISO 22000 was first introduced in 2005 by the International Organization for Standardization, and a few weeks back, ISO has released the first revision after almost 13 years of successful implementation of the first edition. The new edition has created a stronger emphasis on the responsibility of leadership, and it applies a risk-based approach to the development of the food safety management systems (FSMS) while eliminating or minimizing most of the concerns raised by many interested parties.

The new version has adopted the common platform of the ISO, which is called High Level Structure or Annex SL, to reduce the complexities of margining several ISO standards in a single production system. The Annex SL has improved compatibility through the realignment of the structure. Revision has also clarified the key concepts such as Plan-Do-Check-Act cycle and risk base thinking by further segregating them into two separate interconnected cycles where one cycle is covering the management system, and the other is covering the principles of HACCP and risk assessment, but both were delivering a more enhanced single output. Standard has further differentiated the key terms, including critical control points (CCPs), operational prerequisite programs (OPRPs), etc., but still leaving some ambiguous terms amid the new clarification, which had more complexities in the first edition. The standard has also changed the key term HACCP plan in the first edition to hazard control plan while keeping the basics unaffected.

However, improvements to the standard have created additional works for the quality assurance personnel operating in the food industry. Since the changes have proposed major differences to the existing ISO 22000 FSMS and the way of presenting the documented information for the propose of system compliances, following generic model has tried to address them in a systematic approach while reducing the number of documents and systems inside a production facility. The model was developed based on an orthodox black tea manufacturing facility which can be used as an externally developed FSMS while adapting to any specific facility with minimum changes. However, it is a generic model where it may not be extremely comprehensive to comply with all the specific requirements. Hence, modifications of the generic model according to individual factory requirements are mandatory with the help of an expert. The objective of the document is to explain the development process of the new standard through a practical example which also can be directly converted into an FSMS in real-time. Nonetheless, the generic model also can be used as a primary document by students, QA personnel, consultants, any interested parties on the ISO 22000:2018 to develop any FSMS for other respective food items where users need to conduct a complete hazard analysis.

The generic model has 70 different documents covering level 1, 2, 3, and 4 of the documentation pyramid, including scope, food safety policy, objectives, manual, food safety procedures, prerequisite programs, hazard analysis, work instructions, checklists, reports, records, validation, verification, and guides, etc. However, it will not provide specific guidance of preparation in each section, but a user guide is provided to assist the user through the modification process.

The tea industry was selected since it is the second largest beverage consumed in the world and the orthodox format generates a number of varieties based on centuries-old production techniques. Nonetheless, most of the systems are paper-based systems while major employees are poor rural mass who lives under the poverty line in most of the tea growing economies in the world. Hence, tea manufacturing mostly employs few knowledgeable people in their production systems due to the low margins entertained and highly labour intensive nature of the tea growing and manufacturing process. Thus, the generic model may leave them an option to accesses a complete set of documents if they are interested in implementing a real system. The system also intends to serve the students in the universities and colleges who are considering a carrier in the food trade who may not have access to a complete food safety system for their study because most of the study projects they are involved in their internships, they might have a requirement to study the FSMS or they are willing to do so, but the existing food safety systems in an operating environment do not allow students to completely studying due to its own traits of confidentiality terms. At the same time, the book ISO 22000:2018 Generic Model trying to offer an option for people who has a basic understanding or willingness to learn ISO 22000 FSMS a practical example with more interactive learning by modifying option.

The additional development support will be published on the ISO 22000 Resource Center as and when necessary because it was designed to read the standard and find out information related to your industry to fill out given formats, instead of considering how and what to do after reading the standard. Hence users are instructed to ask questions themselves and to do it yourself.

USER GUIDE

The guide to customization of externally developed food safety system

1. Hire a consultant who has extensive experience in tea manufacturing and quality assurance systems development in the tea manufacturing industry.
2. Expose the food safety team to a series of external training that guide them to develop and understand the food safety system of the company.
3. Purchase the ISO 22000:2018 version from the standard institution of your country.
4. Check with your country's tea manufacturing governing organization and standard institutions for their industry-specific guidelines and local standards and buy them as reference materials.
5. Collect information on food safety issues and risk concerns regards to your country's tea industry and collect them.
6. Train internal auditors to evaluate food safety systems and continuously build a food safety culture inside your organization.
7. Conduct brainstorming sessions, study, and discuss the generic model within the food safety team before start working on it.
8. Prepare a list of things you need in order to complete the system development, including human and physical resources to complete the task at hand.
9. Study the gap analysis methods discussed in the ISO 22000 Resource Center blog on google, which has extensive coverage of all the areas under your consideration. Then conduct an extensive gap analysis for both existing systems and brand new systems planned.
10. If you need any additional support, you can reach me at vindikal@gmail.com or post a comment on the blog, which will be immediately answered.
11. To download the training presentations specific for tea production, please send me a mail to send the link to the download, or you can download from the same website, which will be posted in the research paper repository of the blog.
12. Evaluate physical requirements under the infrastructure and documentation requirements under system development.
13. Conduct a management review and discuss strategies for the physical developments as well as documentation requirements while developing an execution plan with a time schedule including target deadlines, responsibilities, and expected outputs.

14. Modify the given system while changing appropriate designations, responsibilities, and company-specific data, which are mostly given in the blue coloured texts and the company logo. However, the consultant and food safety team should thoroughly study the generic model while carefully reading all the text where other additional specific changes related to the system can be determined through this exercise.

15. Use the existing hazard analysis, but add or remove additional findings specific to your factory environment with the help of the consultant hired. On the other hand, given hazard analysis is only an example with most commonly considered food safety risks, but it is the user's responsibility to consider facility-specific risk factors for its products.

16. Prepare additional risk profiles if necessary, draw your own specific process flowcharts, site schematics, and specific SOPs where necessary.

17. Most of the checklists, guidelines, work instructions, procedures, PRPs already developed and added to the generic model after careful study of a large number of tea processing facilities, but there are various other additional documents that may be required according to your mock audit before the certification, which has to be developed by the user based on existing document formats. In case anyone required additional support in such instances, contact through the given e-mail without delay.

18. Hire a professional pest controller and develop an individual bait plan concerning your facility based on the pest control threats identified.

19. The system also includes a food defence procedure that can also be modified according to the individual requirements, which has covered most of the industry's risks considered.

20. Instead of developing a whole new system for your production facility, try to amalgamate existing documents to the system while providing reference numbers and modifying existing records to facilitate additional record requirements, which is easier to adapt.

21. The system is already merged with the 5S requirement, where most of the documents do not require additional data to maintain a well-implemented 5S system.

22. The system also has the flexibility to run with quality assurance systems such as ISO 9001:2015, ISO 14001:2015, where few additional documents are required to be added while customizing the existing system to deliver preferred requirements. The given system is already tested for such operations.

23. There are no specific guidelines provided for each section within the system because the guideline documents are ISO 22000 family of standards, where you have to study yourself the ISO 22000:2018 while using the support of a consultant who has expertise in the field to get training and customizing work.

24. However, it is not a very difficult task to modify the generic model by any individual who has a good understanding of the food safety and ISO 22000:2005 ((the previous edition) standard. If you want to understand any topic or any specific activity such as hazard analysis, simply use the online resources where you will find enough information whatever your issue is. Nevertheless, most of such educational information is available on my blog page, ISO 22000 Resource Center, but if you need any additional exploration on any topic, please let me know.

25. The given system has been developed based on ISO 22000:2018, food safety management system requirements. However, it has most of the requirements to comply with FSSC 22000, where users can add few additional requirements as required by FSSC 22000 for the ISO 22000:2018 requirements if you are intended for further certifications such as FSSC 22000. Then facility can enjoy GMP, HACCP, ISO 22000, and FSSC 22000, which cover almost all the global food safety requirements.

26. The given generic model is written based on the medium-risk orthodox black tea manufacturing process. But most of the procedures, manual, and data collection sheets can be used for any other industry with minimum changes to the content. However, individual hazard analysis has to be conducted specifically to the industry while using the given formats.

27. The word copy of the system, which you can easily modify, can be purchased individually by requesting through my personal mail.

I wish you all the best in adapting the revisions of ISO 22000:2018 and developing your food safety management system. If you need any assistance, please drop a mail at vindikal@gamil.com, where you will have immediate assistance 24/7.

XYZ Tea Factory

ISO 22000:2018 Food Safety Management System

XYZ TEA FACTORY
ISO 22000 FOOD SAFETY MANAGEMENT SYSTEMS
ORTHODOX BLACK TEA MANUFACTURE

SCOPE

The ISO 22000 Food Safety Management System covers the activities related to orthodox black tea manufacturing, packaging, and storing at XYZ Tea Factory. Collection of brought green leaf from out-growers and the delivery of processed black tea up to auctioneer's stores will also be covered.

All pack sizes of all grades are included in the ISO 22000 Food Safety Management System.

……………………………………..

Director/ General Manager

XYZ TEA FACTORY
ISO 22000 FOOD SAFETY MANAGEMENT SYSTEM
ORTHODOX BLACK TEA MANUFACTURE

FOOD SAFETY POLICY

We will make sure to provide our customers hygienically safe black tea and added value to their teacup.

We will ensure to provide pure orthodox black tea made at our factory is free from physical, chemical, and microbiological hazards which are conforming to the relevant statutory and regulatory requirements.

We will train and motivate our employees continuously to upgrade and maintain ISO 22000 FSMS effectively.

……………………………………

Director/ General Manager

XYZ TEA FACTORY – ABC GROUP PRIVATE LIMITED

ISO 22000 FOOD SAFETY MANUAL – ORTHODOX BLACK TEA

COPY NO	0	X	MASTER COPY
			CONTROLLED COPY
			AUDIT COPY

....

FOOD SAFETY TEAM LEADER

DATE ISSUED: 01 – 11 – 2018

XYZ TEA FACTORY
ABC Group Tea Factories

FOOD SAFETY MANUAL
Orthodox Black Tea Manufacture

....
Reviewed By
Consultant – FSMS

....
Approved By
General Manager/Director

Table of Content

AMENDMENT/REVISION RECORD SHEET

The food safety team leader is responsible for maintaining this Amendment/Revision Record Sheet.

Introduction or Clause No	Page Amended	Date Amended	Date Issued (Revision)	Revision Number	Description	Approval

DEFINITIONS AND ABBREVIATIONS

Definitions

CUSTOMER;

The distributor (wholesale trader) who purchases bulked orthodox black tea from XYZ Tea Factory at the Colombo Tea Auction through DEF Brokers (Pvt) Ltd, or other various auctioneers. The distributor, in turn, markets the products to retail outlets or institutions in the international market as well as local markets at their own different trade names or brands.

Abbreviations (in alphabetical order)

AFO	Assistant Factory Officer
C - FSMS	Consultant - Food Safety Management System
CCP	Critical Control Point
CL	Checklist
D. Ref	Document Reference
D/GM	Director/General Manager
F	Format/Form
FR	Firing Room
FSMS	Food Safety Management System
FSP	Food safety procedure
FST	Food Safety Team
FSTL	Food Safety Team Leader
GI	Guide
GL	Green Leaf
HA	Hazard Analysis
HACCP	Hazards Analysis Critical Control Point
HC	Hazard Control
HI	Hazard Identification
ISO	International Standardization Organization
OPRP	Operation Perquisite Programmes
P	Prevalence
PL	Plan
PR	Packing Room
PRP	Perquisite Programmes
RM	Raw Material
RP	Report
RP	Risk Profile
RR	Rolling Room
S	Severity
SFO	Senior Factory Officer
SR/GR	Sifting Room/Grading Room
SV	System Validation
TP	Test Procedure
WA	Withering Area
WI	Work Instructions

COMPANY PROFILE

XYZ (Pvt) Ltd is one of the pioneering and oldest orthodox black tea manufacturers in Sri Lanka, which was established in 1896 as a family business. Since then, the organization has been continuously in business manufacturing orthodox black tea. The company is located in the XXXXX, about 17km south of XXXX, one of the major low-grown districts, where conventional orthodox black tea manufacturing is one of the main industries for a long time.

The organization's products are distributed through Colombo tea Auction mainly by DEF Brokers Ltd. and other brokering companies to the local and international markets. The company has implemented 5S in the factory. Company is one of the leading orthodox manufacturers with district rank lees than 5 in most of the time in the year for high sales value.

ACHIEVEMENTS

- In 1922, the Company was awarded the Silver Medal at all Ceylon industrial exhibitions held in Colombo.
- In 1924, the Company received a Diploma issued by the British Empire Exhibition of that year in appreciation of the service rendered to the tea industry.
- In 1991, XYZ (Pvt) Ltd was incorporated to handle the production of orthodox black tea
- In 1996, the Company won the "Entrepreneur of the year" Bronze award conducted by the Federation of Chamber of Commerce, industry of Sri Lanka.
- In 2001, the Company won a merit award in the competition conducted by the Federation of Chamber Commerce, industry of Sri Lanka.

OWNERSHIP

This document is a property of XYZ Tea Factory - ABC Group, XXXXX Road, XXXXXX, Galle, Sri Lanka. This should not be copied or divulged in whole or in part to any party without the prior written authority of the Director/General Manager XYZ Tea Factory - ABC Group Private Limited.

....
Director/General Manager

LOCATION OF SITES

Location of Site

Head office;

ABC Group Private Limited
TRZ Tea Factory,
XXXXXX Road,
Galle,
Sri Lanka.

Telephone : 0094 91 220000 (General)
Facsimile : 0094 91 220000
Email : xyzteafctory@yahoo.com

The factory for the core business of the company is located at;

XYZ Tea Factory
XXXXXX Road,
XXXXXXXX,
Galle,
Sri Lanka.

Telephone : 0094 91 2200000 (General)
Facsimile : 0094 91 2200000
Email : xyzteafctory@yahoo.com
Web : www.xyzteafactory.lk

Chairman:

Land Line : 0094 91 2200000 (direct)
Mobile : 0094 77 7700000

General Manager:

Land Line : 0094 91 2200000 (direct)
Mobile : 0094 77 7700000

Factory Manager

Land Line : 0094 91 2200000 (direct)
Mobile : 0094 77 7700000

Food Safety Team Leader

Land Line : 0094 91 2200000 (direct)
Mobile : 0094 77 7700000

LAND, BUILDING, AND HUMAN RESOURCES

(For information only, not subject to assessment change)

ABC Group is a limited liability company, and XYZ Tea Factory is a subsidiary of the group. The head office of the company is located at;

ABC Group Private Limited
TRZ Tea Factory,
XXXXX Road,
Galle,
Sri Lanka.

The factory for the core business of the company (XYZ Tea Factory) is located at;

XYZ Tea Factory
XXXXX Road,
XXXXXXX,
Galle,
Sri Lanka.

Facilities at XYZ Tea Factory

	Sq. meters
Land area (3 acres)	12,000
Buildings;	
General Administration, Finance	20
Management/Quality Assurance	10
Withering,	1066
Rolling and Fermentation	305
Firing	177
Grading/Sifting	353
Packing	159
Bag Stores	25
Services and miscellaneous units	68
Canteen and other welfare facilities	500
Officer's Rest House	153
Seminar Hall	45
Burner Yard	159

INTRODUCTION - PLANTS AND EQUIPMENT

(For information only, not subject to assessment change)

At XYZ Tea Factory

Receiving Bay
Weighing scales 01 nos
Lift 01 nos

Withering Area
Troughs 08 nos of 25m x 1.8m
Weighing scales 02 nos

Rolling Room & Fermenting Area
Orthodox Rolls 07 nos of 47"
Roll Breakers 04 nos
Vibrators 04 nos
Conveyers 12 nos
Humidifier System 01 nos

Firing
Drier 02 nos
3T fibre sorter 01 nos
Fibre Mat 01 nos
Weighing scales 01 nos

Sifting
Middleton sifter 04 nos
Michi sifter 10 nos
Winnowers 02 nos
Chota (rotary) sifter 01 nos
Cutter 01 nos
Colour separator 03 nos

Packing Area
Weighing scales 02 nos
Moisture meter 01 nos
Sealer 01 nos

Compressor 01 nos situated outside the plant with a distribution system.
Generator 01 nos and firewood reaper 01 nos are located outside the plant area.

LIST OF PROCESSES

1. Collection and withering
2. Rolling and fermentation
3. Firing, sifting, winnowing, and packing
4. Cleaning on the go
5. Control of documents
6. Control of Monitoring & Measuring Equipment
7. Control of nonconforming products
8. Control of food safety records
9. Corrective Action
10. Evaluation of suppliers
11. Human Resources Development
12. Identification of Customer Needs
13. Internal Audit
14. Management Review
15. Preventive Action
16. Preventive Maintenance
17. Production Planning
18. Purchasing
19. Raw/Packaging Material Acceptance
20. Raw/Packaging Material Storing
21. Transporting
22. Water treatment
23. Product Identification and Traceability
24. Monitoring & measurement of process
25. Monitoring & measurement of the product

PROCESS MAP

Pure Orthodox Black Tea

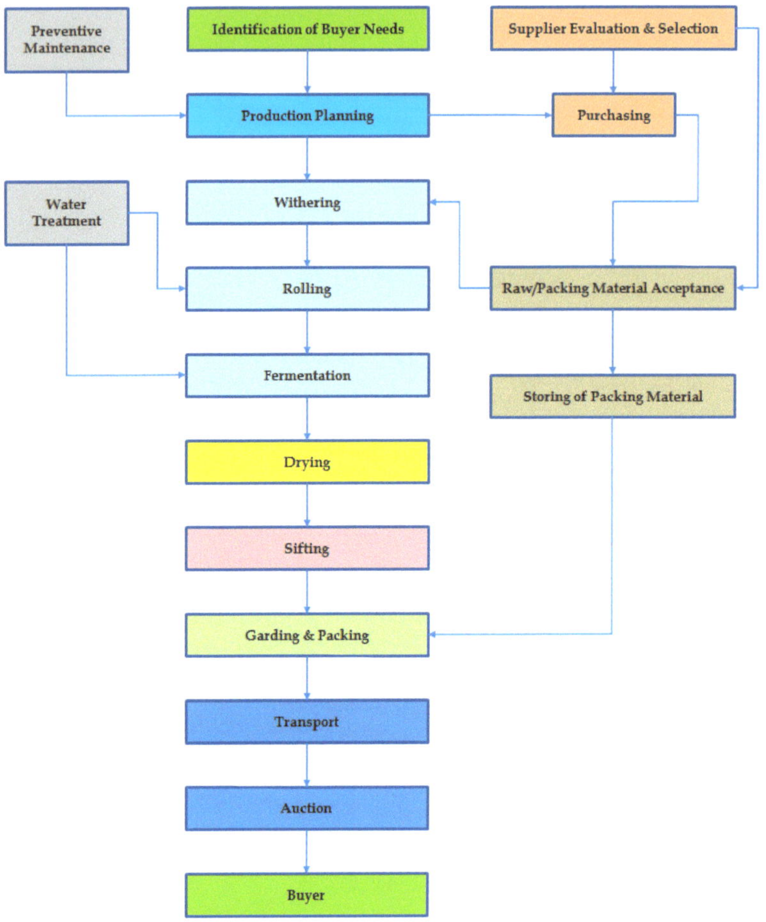

- All calibration activities are connected to raw/packaging material acceptance, withering, rolling, fermentation, firing, sifting, grading, packing, storing, and transportation.
- Process of management review and human resources development are connected to all the above processes.
- Internal audit, corrective action, preventive action processes are connected to all processes.
- Product identification and traceability are connected to raw and packaging material acceptance, withering, rolling, fermentation, firing, sifting, grading, packing, storing, and transport.
- Monitoring and measurement of the process are connected to all production processes.
- Monitoring and measurement of products are connected to raw and packaging material, rolling, fermentation, firing, packing, and storing of the finished product.

MAIN ORGANIZATIONAL CHART

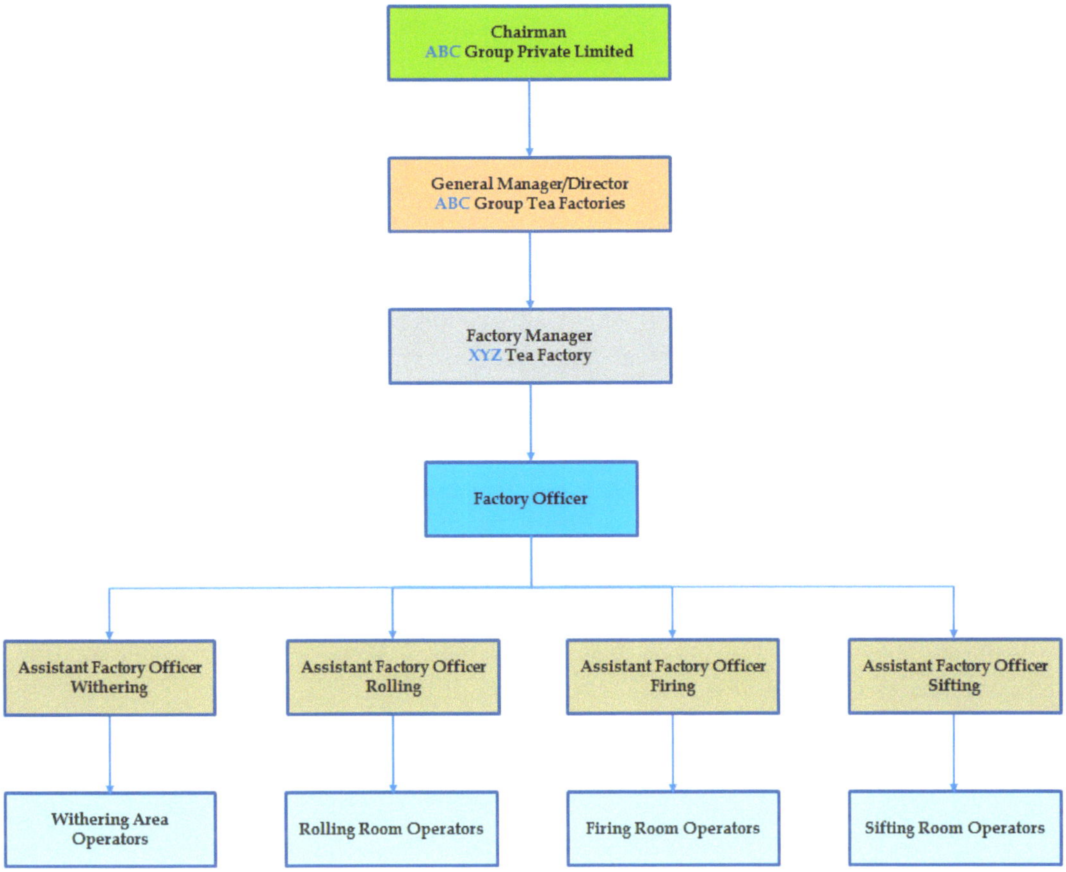

CONTEXT OF ORGANIZATION
(Ref: Clause 4.1 of ISO 22000:2018)

M/s XYZ Ltd. is an orthodox back tea manufacturer and an exporter who is involved in the collection of brought green leaf from out-growers, orthodox black tea manufacturing, packaging, and storing at XYZ Tea Factory, and delivery of processed black tea up to auctioneer's stores.

The top management shall be in charge of the identification of interested parties, and the food safety team is responsible for documenting all the interested parties and their legal, regulatory, contractual, and other requirements and interests. The FSTL shall be responsible for updating the information and distributing it among the FST at least once a year.

The context of the M/s XYZ Ltd. shall consider; the external and internal issues, but not limited to legal, technological, competitive, market, cultural, social, economic environments, cybersecurity, and food fraud, food defence and intentional contamination, knowledge, and performance of the organization, whether international, national, regional or local considerations throughout the manufacturing process including forward and backward linkages, that are relevant to the purpose and that affect company's ability to achieve the intended outcome(s)/results of this food safety management system. The organization shall identify, review and update information related to these external and internal issues.

Hence, all relevant food safety issues shall be considered during the hazard identification and analysis process, which will be evaluated, documented, and updated annually or as and when required by the food safety team.

INTERESTED PARTIES

(Ref: Clause 4.2 of ISO 22000:2018)

The M/s XYZ Ltd., an orthodox black tea manufacturer as to the scope of the processing operation, is considering the following groups as interested parties for the given products to ensure that the organization has the ability to consistently provide products and services that meet applicable statutory/regulatory and customer requirements with regard to food safety.

The organization shall determine:

a. The interested parties that are relevant to the food safety management system;
b. The relevant requirements of the interested parties that are relevant to the food safety management system.

The management is considering stakeholders, persons, or organizations that can influence our food safety/food quality, business continuity, and persons or organizations that can be affected by our food safety/food quality or business continuity activities. Thus, interested parties could include:

1. Customers
2. Employees
3. Suppliers
4. Investors
5. Media
6. Competitors
7. Others

M/s XYZ Ltd. shall establish a documented procedure to address requirements, applications, and compliances to be executed within the organization for the needs and the expectations of interested parties.

Reference

1. Food safety procedure for statutory, regulatory, and customer requirements – FSP 02

SCOPE

(Ref: Clause 4.3 of ISO 22000:2018)

The ISO 22000 Food Safety Management System (FSMS) covers the activities starting from the receiving of green leaves from supplier/field, delivery, withering, rolling, fermentation, drying shifting, blending, and other minor operations related to orthodox black tea manufacturing, packaging and storing at XYZ Tea factory located at XXXXXX, XXXXXXX, Galle. Collection of green leaf from out-growers and the delivery of processed black tea up to auctioneer's stores will also be covered by the FSMS. The food safety issues and requirements relevant to the interested parties are also covered in this food safety management system.

All pack sizes of all grades are included in the ISO 22000 Food Safety Management System.

The products marketed, applicable to the food safety management system are given below;

Made Tea in the following varieties:

Main Grades

BOP – Broken Orange Pekoe
BOP SP– Broken Orange Pekoe Special
BOP 1 – Broken Orange Pekoe 1
BOPF – Broken Orange Pekoe Fannings
FBOP – Flowery Broken Orange Pekoe
FBOPF – Flowery Broken Orange Pekoe Fannings
FBOPF EXSP – Flowery Broken Orange Pekoe Fannings Extra Special
FBOPF EXSP 1 – Flowery Broken Orange Pekoe Fannings Extra Special 1
FBOPF SP – Flowery Broken Orange Pekoe Fannings Special
FBOPF 1 – Flowery Broken Orange Pekoe Fannings 1
FBOP 1 – Flowery Broken Orange Pekoe 1
OP – Orange Pekoe
OPA – Orange Pekoe A
OP 1 – Orange Pekoe 1
PK – Pekoe
PK 1 – Pekoe 1
Silver Tips
Golden Tips

Off Grades

BOP 1A – Broken Orange Pekoe 1
BM – Broken Mix
FNGS – Fannings
Dust

FOOD SAFETY MANAGEMENT SYSTEM

(Ref: Clause 4.4 of ISO 22000:2018)

Scope

The scope covers all operations and activities within the food safety management system. The manual describes the relevancy to the ISO 22000:2018 standard and exact documented reference for each mandatory requirement within specific food safety measures planned and implemented by XYZ Ltd.

General requirements

M/s XYZ Ltd. has established, documented, implemented, operating, and updating a food safety management system aimed at providing products that, according to their intended use, are safe for consumers and continually improve their effectiveness in accordance with the requirements of ISO 22000:2018 standard.

The organization has

a) Determined the process needed for the food safety management system and its application throughout the organization to demonstrate compliance with applicable statutory/regulatory food safety requirements;
b) Determined the sequence and identification of the process to ensure that the organization conforms to its stated food safety policy;
c) Determined the criteria and methods needed to evaluate and assess food safety customer requirements and demonstrate conformity with those mutually agreed upon customer requirements that relate to food safety;
d) Determined the specific people and resources to effectively communicate food safety issues to interested parties within the food chain;
e) Ensure that both the operation and control of these processes are safe and effective to demonstrate conformity to relevant interested parties;
f) Ensured the availability of resources and information necessary to support the operation and monitoring of these processes;
g) Ensure these processes are monitored, measured, and analysed and;
h) Implement action necessary to achieve planned results and continual improvement of these processes.

These processes have been managed by the top management in accordance with the requirements of the ISO 22000: 2018 standard.

The food safety management system shall be established, documented, implemented, and maintained to fulfill the requirements of ISO 22000:2018 standard to assure the products listed in scope are free of biological, chemical, and physical hazards. This system shall be reviewed and updated when and where necessary.

Reference

1. ISO 22000:2018 manual
2. Documented food safety producers
3. Hazard control plan
4. Operational prerequisite programmes
5. Prerequisite programmes
6. Work instructions

LEADERSHIP

(Ref: Clause 5.1 of ISO 22000:2018)

Scope

The scope covers the top management responsibilities and their commitment related to activities defined in the food safety management system.

Activities

M/s XYZ Ltd. shall establish the food safety policy as per the context of organization, requirements, and issues of interested parties to comply with the strategic direction of the organization while ensuring the smooth integration of food safety management requirements into the business process.

The food safety policy appropriate to the organization shall be established, communicated, and reviewed when necessary. The objectives that align with the food safety policy shall be established, communicated, and monitored.

Resources required for the effective operation of the food safety management system shall be provided to effectively implement, communicate and comply with statutory/regulatory requirements and mutually agreed upon customer requirements related to food safety.

Management reviews shall be conducted to ensure that the food safety management system is evaluated and maintained to achieve its intended results, and the food safety management system shall be upgraded.

The management shall provide the required strategic direction and relevant human resources to contribute to the effectiveness of the food safety management system while promoting continual improvement.

Top management shall support the rest of the management team to demonstrate their food safety leadership as it applies to their areas of responsibility.

Reference

2. Food safety policy & objectives
3. Food safety procedure for management reviews – FSP 10

FOOD SAFETY POLICY

(Ref: Clause 5.2 and 6.2 of ISO 22000:2018)

We will make sure to provide our customers hygienically safe black tea and added value to their teacup.

We will ensure to provide pure orthodox black tea made at our factory is free from physical, chemical, and microbiological hazards which are conforming to the relevant statutory and regulatory requirements.

We will train and motivate our employees continuously to upgrade and maintain ISO 22000 FSMS effectively.

....

Director/General Manager

FOOD SAFETY OBJECTIVES

1. Reduction of moisture increases up to 50% throughout sifting and packing operations by the end of May 2019.
2. Improve the food safety awareness and competencies among the workforce by the end of May 2019.
3. All executives in charge of raw material, processing, and finished products will be exposed to external training on basic food hygiene and/or ISO 22000 by the end of May 2019.
4. The number of customer complaints due to food safety issues will be reduced by 50% by the end of May 2019.
5. Elimination of the birds from the factory site by the end of December 2019.

The food safety policy and the objectives shall be available to the employees, customers, and relevant interested parties as appropriate, which will be maintained as documented information and communicated, understood and applied at all levels within the organization.

ORGANIZATIONAL ROLES, RESPONSIBILITIES, AND AUTHORITIES

(Ref: Clause 5.3 of ISO 22000:2018)

M/s XYZ Ltd. has defined each individual operator's roles, responsibilities, and the authorities in their hiring agreement, including the job description to make sure smooth operation of the manufacturing process and food safety management system to facilitate intended product quality and consumer safety.

In addition, the responsibilities and authorities are assigned to; ensure conformity of food safety management system requirements, report on the performance of the system to top management, and also to initiate actions through designated persons with defined responsibility and authority.

M/s XYZ Ltd. has nominated the food safety team and appointed a food safety team leader. The team leader shall be responsible for ensuring the food safety management system is established, implemented, maintained, and updated while managing and organising the work of the food safety team and ensuring relevant training of and competencies for the food safety team while reporting to the top management on the effectiveness and suitability of the food safety management system.

Director/ General Manager

1. Commitment to development, implementation, and continual improvement of the effectiveness of FSMS.
2. Define, document, communicate and review the food safety policy.
3. Ensure operation of effective FSMS through guidance & providing resources.
4. Resource allocation for food safety related improvements.
5. Functioning as the chairperson of the management review meeting.
6. Ensure control of the handling of potentially unsafe products.
7. Ensure correction & corrective action.

Factory Manager

1. Function as the food safety team member.
2. Ensure required resources are approved and allocated.
3. Ensure provision of resources for the proper hygienic environment, proper handling, and storage of finished products.
4. Make sure adequate supply of utilities for smooth functioning of production activities and cleaning and sanitation activities of the factory.
5. Maintaining the factory buildings and layouts in an adequate state of repair to prevent possible cross-contamination of food material.
6. Ensure handling, transportation, and storage of finished products.
7. Ensure control of monitoring & measuring equipment.
8. Ensure control of non-conforming products.

Production Manager/Factory Officer

1. Function as the food safety team member.
2. Ensure proper hygienic environment, handling & storage of finished products.
3. Responsible for handling, transportation, and storage of finished products.
4. Responsible for control of monitoring & measuring equipment.
5. Responsible for controlling the non-conforming products.

6. Responsible for control of the handling of potentially unsafe products.
7. Implementation of correction & corrective action.
8. Implementation of control of non-conforming products.
9. Implementation of emergency preparedness procedures.
10. Handling of potentially unsafe products.
11. Make sure all products are manufactured in compliance with set standards.
12. Function as a member of the management review committee.
13. Relevant food safety information is communicated to suppliers and vice versa.

Q A Manager/Quality Officer – Quality Assurance/Food Safety

1. Function as the food safety team leader.
2. Ensure the food safety management system is established, implemented, maintained, and updated.
3. Managing and organizing the work of the food safety team.
4. Ensure relevant training and competencies for the food safety team (7.2).
5. Report to the top management on the effectiveness and suitability of the food safety management system.
6. Planning & implementation of preventive maintenance.
7. Maintaining all machinery and utensils in the factory in an adequate state of repair for smooth functioning of operation.
8. Make sure all materials purchased are within company specifications.
9. Make sure all green leaves are purchased from registered/approved suppliers.

Assistant Factory Officers

1. Maintain adequate personnel hygiene facilities during the shift.
2. Maintain an adequate level of personnel cleanliness and food safety related behaviour in all employees.
3. Maintain process parameters within limits and record keeping.
4. Maintain adequate general cleanliness within and around the production premises.
5. Maintaining the FIFO system for all materials.
6. Maintain an adequate level of temperature and humidity control in production.

Process Operators

1. Accomplish adequate personnel hygiene and facility hygiene during the shift.
2. Accomplish adequate level of personnel cleanliness and behaviour in all operations.
3. Accomplish process parameters within limits and record keeping.

Consultant – Food Safety Management System

1. Establishing, documenting, and implementing FSMS.
2. Implementation of correction and corrective action procedures.
3. Review and update food safety standards of the factory.
4. Planning and implementation of internal food safety audits.
5. Responsible for control of documents and control of records.
6. Function as a member of the management review meetings.

FOOD SAFETY TEAM

(Ref: Clause 5.5 & 7.3.2 of ISO 22000:2018)

ISO 22000 Food Safety Team

Mr. M.R.M Ralph (**Over Roll Leader**)	– Director/General Manager
Mr. M.M.M. Silva	– Factory Manager
Mr. M. W. Pereira (**FST Leader**)	– Quality Officer
Mr. M.M. S. P. James	– Factory Officer
Mr. S. V. Roland	– Assistant Factory Officer
Mrs. M. K. Ana	– Assistant Factory Officer
Miss. U. S. Rebecca	– Assistant Factory Officer
Mr. D. R. Leonard	– Assistant Factory Officer

FOOD SAFETY MANAGEMENT SYSTEM PLANNING

(Ref: Clause 6.1, 6.2, 6.3 of ISO 22000:2018)

Scope

All operations within the food safety management system

Activities

1. All operations within the scope defined shall be planned in a way that the hazards possible to occur in the products are controlled.
2. The FSMS in compliance with ISO 22000 shall be established to fulfill the above requirement and to address risk and the opportunities generated through the manufacturing process.
3. Use the food safety procedure for statutory, regulatory, and customer requirements as the base to collect information to comply with the requirements of interested parties.
4. The risk and the opportunities shall be addressed through enhancing the desirable effects to assure the FSMS can achieve its intended outcome(s)/results while preventing or reducing undesired effects and achieve continual improvement.
5. M/s XYZ Ltd. shall plan its actions to address identified risks and opportunities while integrating and implementing the actions into its FSMS processes, which can evaluate the effectiveness of those actions.
6. The system and manual are developed based on the ISO 22000:2018, CODEX GAP, GHP, GMP Principles, ISO 9001:2015, ISO 3720:2011, 5S, SLS 135, Sri Lanka Tea Board Circular No.: AL/MQS-Rev/2010 26th April 2010 by the SLTB with Annexure - A for the Sri Lankan Origin teas.
7. The actions taken shall be proportionate to the potential impact on food safety requirements, and the conformity of food products and services to customers, and the requirements of interested parties in the food chain.
8. M/s XYZ Ltd. has established food safety objectives (clause 6.2.1) which shall be consistent with the food safety policy that is measurable and consistent with applicable food safety requirements, including statutory/regulatory and customer requirements, and it shall be implemented at relevant functions and levels.
9. The aforesaid objectives of the food safety management system have been developed according to clause 6.2 of the ISO 22000:2018, which is consistent with the food safety policy and measurable where practicable.
10. The specific customer requirements shall be considered case by case, and relevant compliances will be covered accordingly.
11. Periodical audits, management reviews, and updating activities shall be conducted to monitor and verify the system compliances and communicated to relevant interested parties.
12. The system shall maintain, update and verify as appropriate, where the company shall retain documented information on the objectives of the food safety management system.
13. When planning how to achieve its objectives of the food safety management system, the company has determined:
    a. What will be done shall be defined through the FSPs, PRPs, hazard analysis process, and WIs documented to execute the FSMS;

b. What resources will be required has been defined throughout the processes explained in the relevant process, and physical resource requirement shall be evaluated and documented while conduction gap analysis process or while re-validating PRP 01 requirements;
c. The responsibilities have been defined as appropriate and shall be modified change as requirements arise throughout the operations;
d. The entire implementation shall be completed within six (6) months of the initial management review meeting, and the second management review meeting shall be conducted to review the implementation process and review the results;
e. The results will be evaluated through internal audits, mock audits, mock recalls, validation and verification activities, as well as statistical methods based on the collected data after operating the system at full compliance.

14. Any change which can have an impact on food safety shall be communicated to the food safety team. The food safety team shall evaluate and approve the change before implementation.
15. M/s XYZ Ltd. has determined the feasible methods to follow when changes are required for the food safety management system, including personnel changes that shall be carried out and communicated in a planned manner.
16. The FST shall determine the need for changes, including personnel, processes, or FSMS, with consideration to potential consequences for the supply and maintenance of safe food production, the integrity of the FSMS, the availability of resources to effectively implement the changes and allocation or re-allocation of responsibilities and authorities.

References
1. ISO 22000 manual
2. Food safety procedure for statutory, regulatory, and customer requirements
3. Documented producers
4. Hazard control plan
5. Operational prerequisite programmes
6. Prerequisite programmes
7. Work instructions

SUPPORT AND RESOURCE MANAGEMENT

(Ref: Clause 7.1 of ISO 22000:2018)

Scope

The scope covers resources required for the management of the human resource, infrastructure, work environment, and externally developed elements or provided processes, products, and services for the effective implementation, operation, maintenance, updating, and continual improvement of the food safety management system.

Activities - General

1. Provide required physical, monetary and human resources for the development, implementation, maintenance, updating, management, and continuous improvements of the system based on the capability and any constraints on existing internal resources as well as resources required from external sources.
2. M/s XYZ Ltd. has provided the resources required for the establishment, implementation, maintenance, updating, and continual improvement of the FSMS with due consideration for capabilities and any constraints of existing resources as well as resources required from external sources.
3. M/s XYZ Ltd. has determined to implement an externally developed ISO 22000:2018 FSMS generic model, which will be customized to match company requirements.

Human Resources (7.1.2)

1. M/s XYZ Ltd. has identified the necessary competencies of those; whose activities have an impact on food safety.
2. M/s XYZ Ltd. has provided competent persons (7.2) that are necessary to operate and maintain an effective food safety management system.
3. The company shall also provide assistance from external experts for the development, implementation, operation, and assessment of the FSMS.
4. FST shall maintain evidence of agreement or contracts defining the competency, responsibility, and authority of external experts are retained as documented information.

Infrastructure (7.1.3)

1. M/s XYZ Ltd. has provided adequate resources for the determination, establishment, and maintenance of the infrastructure facilities necessary to achieve conformity with the requirements of the ISO 22000 and to initiate and operate FSMS effectively.
2. The resources provided shall include but not be limited to land, vessels, buildings, and associated utilities, equipment, including hardware and software, transportation, information, and communication technology.

Work environment (7.1.4)

1. M/s XYZ Ltd. has determined to provide adequate and appropriate resources for the establishment, management, and maintenance of the work environment necessary to achieve conformity with the requirements of the food safety management system.

2. The top management has decided to provide resources to crate positively active socially and psychologically engaged non-discriminatory, emotionally proactive workforce with desired physical resources such as appropriate temperature, humidity, light, airflow, hygiene, and noise in the work environment.
3. All working environments shall be established, managed, and maintained in order to prevent contamination of food.

Externally Developed FSMS/Elements (7.1.5)

1. M/s XYZ Ltd. had obtained this externally developed food safety system through a reputed vendor and appointed a subject expert (food safety consultant) to customize it before it was established.
2. The expert with FST shall customize the system with the food safety team to ensure that the use of externally developed food safety management system or its extracted elements, including PRPs and the hazard control plan in accordance with factory requirements by:
    a. Developing in conformance with requirements of ISO 22000:2018;
    b. The system is applicable to the sites, processes, and products of the organization;
    c. The system is specifically customized and adapted for the hazard analysis process by the food safety team to the processes and products of the M/s XYZ Ltd.; and,
    d. The system shall be implemented, maintained and updated, and continually improves as required by ISO 22000:2018;
3. The initial generic model and the modified/customized system with relevant explanations for modifications shall be retained as documented information.

Control of Externally Provided Processes, Products, or Services

1. The M/s XYZ Ltd. shall establish and apply criteria for the evaluation, selection, monitoring of performance, and re-evaluation of suppliers prior to registration and annually, external providers of processes, products, or services.
2. The FST shall ensure the externally provided processes, products, or services do not adversely affect the ability to consistently meeting the requirements of the FSMS.
3. FSTL shall ensure the adequate communication of the company's requirements to the external provider(s) to ensure that external providers conform to the established criteria.
4. FST shall ensure that external service providers conform to the established criteria (clause 7.1.2).
5. FST shall maintain documented information of these activities and any necessary actions as a result of the evaluations and re-evaluations as evidence.

References

1. Construction, layout and equipment	– PRP 01
2. Preventive maintenance	– PRP 03
3. Personnel hygiene	– PRP 06
4. Sanitary and employee facilities	– PRP 08
5. Work environment	– PRP 10
6. Cleaning and sanitisation of plant, machinery equipment and utensils	– PRP 11

COMPETENCE AND AWARENESS

(Ref: Clause 7.2 & 7.3 of ISO 22000:2018)

Scope

The scope covers competence, awareness, and training of the internal human resources and the external service providers.

Activities

1. FST shall determine the necessary competence of person(s), including external providers working under the company's control which affects its food safety performance and effectiveness of the FSMS.
2. The HR department and FST shall ensure hiring and maintenance of records for all relevant personnel, including the food safety team and those responsible for the operation of the hazard control plan, are competent on the basis of appropriate education, training, or experience.
3. M/s XYZ Ltd. shall ensure its food safety team has a combination of multi-disciplinary knowledge and experience for the development and implementation of the FSMS, including but are not limited to the organization's products, processes, equipment, and food safety hazards.
4. FST shall acquire the necessary competence through hiring or training where applicable and evaluate the effectiveness of the actions taken.
5. Awareness and training shall be provided to fulfill the gaps in expected and existing competencies.
6. Evaluate and record the effectiveness of training.
7. M/s XYZ Ltd. shall ensure all relevant persons who work under their control shall be aware of the food safety policy and the food safety objectives relevant to their task(s).
8. The FST shall be aware of all the individuals and their contribution to the effectiveness of the FSMS, including the benefits of improved food safety performance and implications of not conforming to FSMS requirements.
9. FST shall maintain and retain appropriate documented information as evidence of competence of relevant personnel, including training records and evaluations conducted.

References

1. Personnel information of all relevant persons including education, experience, training, and skills (personnel files)
2. Hiring agreements and job descriptions
3. Contracts and personnel data of the external providers
4. Training records on food safety

COMMUNICATION

(Ref: Clause 5.6 of ISO 22000:2018)

Scope

The scope covers external and internal communication throughout the food chain applicable to the scope defined for the food safety management system.

Activities

1. M/s XYZ Ltd. has established a communication procedure that explains internal and external communication strategies and specific activities relevant to the ISO 22000:2018.
2. M/s XYZ Ltd. has an established mechanism for communicating food safety related issues, including what it will communicate, when to communicate, with whom to communicate, how to communicate, and who communicates within the organization as well as external parties along the food chain.
3. M/s XYZ Ltd. has assigned the food safety team leader as the authorised person for liaison with relevant parties related to food safety.
4. Effective communication strategies are used for effective communication, which is understood by all persons whose activities have an impact on food safety.

External communication

1. M/s XYZ Ltd. has ensured that sufficient information is communicated externally and is available for interested parties of the food chain.
2. The food safety team leader has the authorization to communicate with external bodies such as suppliers, contractors, regulatory and statutory organizations, and distributors, where the content of the communication shall be defined by the General Manager on a case-by-case basis.

Internal communication

1. The M/s XYZ Ltd. has established, implemented, and maintaining effective arrangements for communicating between persons on issues having an impact on food safety.
2. The information and issues related to food safety shall be communicated to all relevant personnel, including the food safety team.
3. This information shall be used for updating the FSMS.

Reference

1. Food Safety procedure for communication – FSP 03
2. Correction and corrective action – FSP 06

DOCUMENTED INFORMATION

(Ref: Clause 7.5 of ISO 22000:2018)

Scope

The scope covers documented information required for the FSMS with their controlling mechanism determined by the organization as being necessary for the effectiveness of the FSMS and food safety requirements required by statutory/regulatory authorities and interested parties.

Activities

1. M/s XYZ Ltd. has established a procedure for control of documented information which explains authorities to view or permission to change, control, retention, distribution, and preservation of information as evidence for reference, traceability, and auditing as specified by ISO 22000:2018.
2. The applicable food safety requirements, including statutory, regulatory, and customer requirements pertaining to the safety of orthodox black tea production, have been defined and document as a food safety procedure.
3. The method of controlling the documented information has been defined and documented with the method of approval, review, and updates for documented information.
4. The method of identification, storage, protection, retrieval, retention, and disposition or obsolete, amend or revise has been defined and documented.
5. A list of controlled copies of documents of internal and external origin shall be maintained.
6. The manual and the procedures shall be reviewed by Consultant - FSMS and approved by General Manager.
7. The PRPs, OPRPs, hazard control plan shall be reviewed and approved by Consultant - FSMS and Factory Manager.
8. All amendments shall be approved by General Manager and be recorded in Amendment/Revision Record Sheet and implemented only after approval.
9. All revisions to any clause of the food safety manual shall be approved by the General Manager and implemented only after the approval.
10. All revisions shall be issued on a new page(s) for every controlled copy in existence. The revision number of all pages shall be increased by a unit with a revised issue date. Details of all revisions shall be recorded in the Amendment/Revision Record Sheet.
11. The distribution of this Food safety manual shall be controlled and approved by the Factory Manager. The distribution list is given hereunder;

Copy No	Designation
Copy 0	Master copy
Copy 1	Audit Copy
Copy 2	Managing Director/Chairman
Copy 3	Director/General Manager
Copy 4	Factory Manager
Copy 5	Factory Officer
Copy 6	Quality Officer
Copy 7	Consultant – FSMS

References

1. Food safety procedure for control of documented information – FSP 01
2. Food safety procedure for statutory, regulatory, and customer requirements – FSP 02
3. List of controlled copies of internal documents

OPERATIONAL PLANNING AND CONTROL

(Ref: Clause 8.1 of ISO 22000:2018)

Scope

The scope covers all the operational, planning, and control activities undertaken to achieve the realization of safe products required by statutory/regulatory authorities and interested parties.

Activities

1. M/s XYZ Ltd. has determined the methods for proper operations (plan, implement, control, maintain and update), which shall be explained in detail through FSPs, PRPs, OPRPs, and hazard control plan according to the requirements of the FSMS based on ISO 22000:2018 standard and other applicable statutory and regulatory requirements.
2. The processes have been designed for the realization of safe products through establishing criteria for the specific processes and implementing control of the processes in accordance with the relevant criteria while keeping documented information to the extent necessary to have the confidence to demonstrate that the processes have been carried out as planned.
3. M/s XYZ Ltd. shall ensure that the outsourced processes are implemented in accordance with planned food safety requirements.
4. The company shall ensure that outsourced processes are controlled according to compliance requirements and verified through regular audits.
5. The company shall control planned changes and review the consequences of unintended changes while taking action to mitigate any adverse effects when and where necessary.

References

1. Documented information
2. Record, revision, and amendment sheet
3. Supplier selection, verification, and audits

PREREQUISITE PROGRAMS
(Ref: Clause 8.2 of ISO 22000:2018)

Scope
The scope covers all basic physical conditions and activities that are necessary to maintain a hygienic environment throughout the food chain and to establish a contamination-free work environment in the factory, storage, and transportation.

Activities
1. The PRPs shall be designed to prevent and/or reduce the likelihood of introducing contaminants (including food safety hazards) to the product, which shall prevent and/or reduce the biological, chemical, and physical contamination of products, including cross-contamination between products.
2. The PRPs shall facilitate the prevention and/or reduction of contaminants (including food safety hazards) in the products and product processing environment.
3. Prerequisite programs appropriate to the nature and the size of the operation shall be established, documented, approved, and implemented by the food safety team through relevant departments.
4. All PRPs are implemented prior to implementation of hazard analysis, monitored, and effectiveness shall be evaluated by the food safety team based on the requirements of clause 8.2.1, 8.2.2 of ISO 22000:2018. The modifications/changes shall be implemented after hazard analysis.
5. M/s XYZ Ltd. shall consider the ISO/TS 220002- 1 food manufacturing, Good Manufacturing Practices, and Good Agricultural Practices as the main reference documents. The following requirements and conditions shall be used when and where necessary while designing its PRPs to comply with statutory/regulatory requirements which are specific to individual PRPs.
   a. construction and layout of buildings and utilities;
   b. the layout of premises, including zoning, workspace, and employee facilities;
   c. supplies of air, water, energy, and other utilities;
   d. pest control, waste and sewage disposal and support services;
   e. the suitability of the equipment and its accessibility for cleaning, maintenance, and preventive maintenance;
   f. supplier selection, approval, and assurance processes (e.g., raw materials, ingredients, chemicals, and packaging);
   g. receiving of incoming materials, storage, dispatch, transportation, and handling of products;
   h. measures for the prevention of cross-contamination;
   i. cleaning and disinfecting;
   j. personal hygiene;
   k. product information/consumer awareness;
   l. others as appropriate
6. M/s XYZ Ltd shall specify selection, establishment, applicable monitoring, and verification procedures in the relevant PRPs and prepare relevant data retrieval formats as documented evidence of the FSMS.

References

1.	Construction, layout and equipment	- PRP 01
2.	Management of purchased materials	- PRP 02
3.	Preventive maintenance	- PRP 03
4.	Control of monitoring and measuring equipment and methods	- PRP 04
5.	Storing, handling and transportation of material	- PRP 05
6.	Personnel hygiene	- PRP 06
7.	Pest control	- PRP 07
8.	Sanitary and employee facilities	- PRP 08
9.	Supply of air, water and energy	- PRP 09
10.	Work environment	- PRP 10
11.	Cleaning and sanitisation of plant, machinery, equipment and utensils	- PRP 11
12.	Supplier evaluation & selection	- PRP 12
13.	Food defence	- PRP 13

TRACEABILITY SYSTEM

(Ref: Clause 8.3 of ISO 22000:2018)

Scope

The scope covers the identification of incoming materials from the suppliers, finished product lots, and relevant batches of raw, in-process, and packing materials of made tea with the initial distribution route.

Activities

1. The traceability system shall be developed in relation to a lot of received materials, ingredients, and intermediate products with relation to the end products and reworking of materials/products and distribution route of the end product as well as compliance with statutory/regulatory and customer requirements.
2. Records shall be maintained as documented information with the details of the date of receipt/batch no. and the name of the supplier.
3. Batches of intermediate or finished products shall maintain all the processing-related or transferring paths as well as specific operators involved throughout the manufacturing process.
4. The above information shall be used as a reference when the above materials are used for production.
5. M/s XYZ Ltd has defined the specific period to retain the documented information as evidence of the traceability system, which shall be a minimum of shelf-life of the end product.
6. The verification of the traceability shall be conducted periodically to verify and test the effectiveness of the traceability system.

Reference

1. Control of documented information – FSP 01
2. Food safety procedure for product identification and traceability – FSP 04
3. Documented information on traceability.

EMERGENCY PREPAREDNESS AND RESPONSE
(Ref: Clause 8.4 of ISO 22000:2018)

Scope
The scope covers the emergency situations and accidents that can have an impact on food safety within the scope of the food safety management system.

Activities
1. M/s XYZ Ltd. has prepared to identify preventive actions for the potential emergency situations and incidents that may have an impact on food safety and which are relevant to our role in the food chain.
2. The potential emergency situation shall be identified considering the environment and the activities of the food chain.
3. The action plans for the crisis situations shall be documented, which are in compliance with statutory/regulatory requirements.
4. Documented information has been established (emergency preparedness and response procedure - FSP 05) and maintained to manage these situations and incidents.
5. Precautionary measures shall be established in order to prevent and/or minimise the impact on products.
6. The planned methods to the emergency situations shall be tested periodically.

Reference
1. Food safety procedure for emergency preparedness and response - FSP 05
2. Emergency control guide - WI 08
3. Documented information on training, mock fire drills, and periodic testing

HAZARD CONTROL

(Ref: Clause 8.5 of ISO 22000:2018)

Scope

The Scope covers planning, provision of resources, implementation, and effective operation of the hazard control plan/HACCP system.

Activities

1. The FST shall collect preliminary information, updated and maintained, which shall include but not be limited to the company's products (orthodox back tea), processes, customer requirements, equipment, all applicable statutory/regulatory food safety requirements (FSP 02), and food safety hazards relevant to the food safety management system.
2. FST shall develop and use the format containing specific requirements given in clause 8.5.1.2 for the raw materials, ingredients, and product contact material. The separate format shall be used containing requirements specified in characteristics of end products (clause 8.5.1.3) and intended use (clause 8.5.1.4) as appropriate.
3. FST shall maintain documented information concerning all raw materials, ingredients, and product contact materials as well as characteristics of end products with respective intended uses to the extent needed to conduct the hazard analysis.
4. The FST shall establish, maintain and update flow diagrams as documented information indicating the sequence and interaction of the steps in operation, any outsourced processes, specific entry points of raw materials, ingredients, processing aids, packaging materials, utilities, and intermediate products that enter the flow, any possible reworking and recycling take place, and where end products, intermediate products, by-products, and waste are released or removed.
5. The FST shall conduct this process for all the products or product categories and the processes covered by the FSMS with onsite confirmation for the accuracy of the flow diagrams, with verification for updates as appropriate and retain as documented information.
6. The FST shall prepare site schematic diagrams showing food and non-food handling areas, processing equipment and contact materials, processing aids, and flow of materials as appropriate and consider potential contaminations while onsite verifications.
7. The plant and equipment (page 10), list of processes (page 11), and process map (page 12) are given at the beginning of the document describing the production process and its amenities, which shall be indicated in the process flow charts and site schematics including any additional requirements indicated by interested parties.
8. The FST shall collect information to conduct the hazard analysis and maintain documented information based on the preliminary information (clause 8.5.1.1 to 8.5.1.5) to determine the hazards that need to be controlled.
9. The hazard analysis shall be conducted based on Codex HACCP principles and ISO 22000:2018 requirements while using the preliminary information collected to determine the hazards that need to be controlled.
10. The FST shall prepare a hazard identification table including all the process steps, raw material, ingredient, product contact material, site schematics, and end products with the hazard identified, their acceptable levels or limits, as well as

where the hazard shall be controlled with reference to specific PRP, CCP or OPRP or combination of control methods applied.

11. The FST shall prepare a document indicating the hazards that are not controlled at the premises, which are the responsibility of specific suppliers with relevance to the specific control measures and supplier verifications applicable.
12. The FST shall conduct the hazard assessment for each identified food safety hazard to determine whether its prevention or reduction to an acceptable level is essential.
13. The FST shall identify potential risks to the production, which can be biological, chemical, or physical, and prepare risk profiles to understand their severity while considering the probability of occurrence in the tea manufacturing process.
14. The FST shall develop appropriate document formats to document information that can be easily retrieved by any interested parties without difficulties understanding. The document formats shall include hazard identification (FSMS 02/RA/HA/HI/01 - 06), hazard assessment (FSMS 02/RA/HA/07 - 11), hazard control plan (FSMS 02/HC/01 - 05) including recording formats.
15. A hazard control plan shall be developed using the following steps/documents.

    - Raw material descriptions
    - End product description and intended use of the product
    - Site schematics
    - Process flow diagrams
    - Hazard identification table
    - Un controllable hazards
    - Hazard analysis table
    - The methodology used for hazard assessment
    - Descriptions of process steps and control measure
    - CCP decision table
    - Summary of hazard control plan including CCP and OPRP plan
    - Monitoring procedures - Hazard control plan
    - Verification procedures

16. The FST shall establish, implement and maintain the hazard control plan where control measures are categorized at CCPs or as OPRPs.
17. The segregation of CCPs and OPRPs shall be carried out through a decision tree model, which was derived based on the requirements specified by ISO 22000:2018 (clause 8.5.2.3 and 8.5.2.4).
18. The decision tree model has given below as a reference to explain the logical sequence used to derive critical control points and operation prerequisite programs as explained in the ISO 22000:2018 standard.
19. The hazard control plan shall be maintained as documented information and shall include the information for each control measure at each CCP or OPRP.
20. The hazard control plan shall be implemented and maintained, and relevant evidence retained as documented information.
21. The FST shall provide evidence that the specified monitoring and measuring methods and equipment in use are adequate for the monitoring and measuring activities related to the PRPs and the hazard control plan.

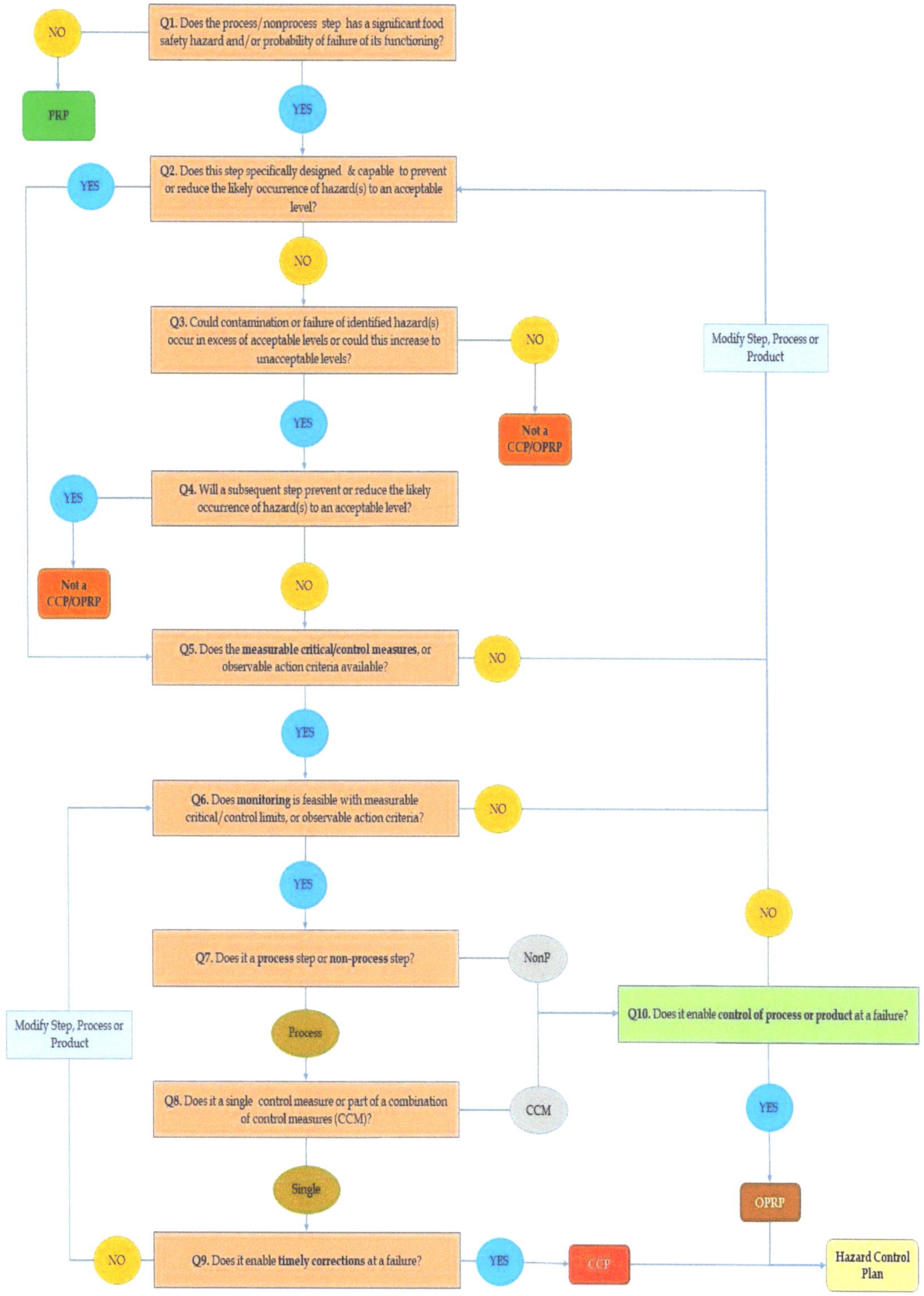

ISO 2200:2018 Decision Tree

22. The segregation of CCPs and OPRPs of ISO 22000:2018 can be achieved using a decision tree, decision table, or a combination of both, and also it can be a matrix, etc. Thus, the following decision tree has been developed based on the specific requirements given in the sub-clauses 8.5.2.4.1 and 8.5.2.4.2. The decision table was designed to answer the same questions to create the documented information.
23. The decision tree was developed considering the simplicity of understanding and easy use, which must be reviewed, amended, and updated if there are any changes to the concept, interpretation of the ISO 22000:2018 standard, or official publication of a decision tree or a table from the ISO/TWC. However, the selection of CCPs, OPRPs, and PRPs are science-based decisions, where the decision tree is optional to improve the logical sequence of the application but not the outcome.
24. Following assumptions were used while developing the decision tree.
    a. If the process step/non-process step has a potentially significant hazard, such steps have to be considered in the risk assessment.
    b. If a control measure does not have a measurable critical limit, it cannot be a CCP as to the technical explanations generally adapted.
    c. If there is no possibility to monitor measurable critical limits or observable action criteria, that control measure cannot be effectively used as a CCP or an OPRP; thus, it needs to be modified and re-evaluated for suitability.
    d. The word "OPRP" re-clarification has clearly defined the current requirements, where a product will not be still injurious to public health at a control failure if there is enough time to make corrections without effectively harming the product. Hence, the control measure cannot be considered as a CCP, which falls under OPRP due to its mild nature of the failure (e.g., a slight increase of moisture levels in made tea).
    e. If the control measure has multiple control measures, it will not be considered as a CCP (e.g., metal detection in made tea).
    f. Thus, a CCP must be a step in the process with a single control measure and measurable critical limits, which enables the correction.
    g. Thus, an OPRP can be a step in the process or not in the process with a control measure, or with a combination of control measures, and with action criteria, or control limits, and measurement, or observations that enable the control of product or process.

Reference

1. Process flowcharts
2. Raw material/ingredient/product contact material descriptions
3. Site schematics
4. End product descriptions
5. Hazard identification
6. Externally controlled/outsourced processes
7. Hazard analysis
8. Hazard control plan
9. Summary of possible foodborne illnesses and related microorganisms
10. Risk profiles and Risk matrix
11. Decision table for segregation of CCPs and OPRPs (FSMS 02/RA/HA/HA/08)
12. Work instructions
13. FSMS reports, records, checklists, and formats

VALIDATION

(Ref: Clause 8.5.3 of ISO 22000:2018)

Scope

The scope covers all control measures for hazards identified through the risk assessment and hazard control plan, including CCPs and operational PRPs.

Activities

1. The FST shall validate control measures prior to implementation, which is to be included in the hazard control plan (clause 8.5.4) and after any change therein (clause 7.4.2, 7.4.3, 10.2, 10.3), for the selected control measures.
2. Each operational PRP and CCP shall be validated to determine if it is capable of achieving the intended control of the identified hazards.
3. The validation shall justify that control measures or combinations of control measures are capable of achieving the intended control of the food safety hazards for which they are designated, and the control measures are effective and capable of, in combination, ensuring control of the identified food safety hazards to obtain end products that meet the defined acceptable levels.
4. The FST shall modify and reassess the control measures or combination of control measures when the validation results show that intended control measures are not effective.
5. Control measures shall be validated using scientific literature, historical knowledge, experimental trials, and data collected during normal operating conditions and statistical applications.
6. The FST shall assess the validity of the previous measurement results when the equipment or process environment is found not in conformity to the set requirements.
7. The FST shall take appropriate action on the equipment or process environment and any product affected.
8. Software used in monitoring and measuring within the FSMS shall be validated by the organization, software supplier, or the third party prior to use, and documented information on validation activities shall be maintained
9. Whenever necessary, validations shall be conducted by external parties. Microbiological or analytical testing shall be used effectively to validate that a process is in control and that an acceptable product is being produced.
10. If additional control measures, new technology or equipment, changes in the control measures, identification of new or emerging hazards or their frequency of occurrence, or unexplained failures of the system occur, re-validation of the system may be necessary.
11. The assessments and resulting action shall be documented.

Reference

1. Validation Guide – FSMS 03/WI/VG/07
2. Handbook on tea manufacture
3. External test reports on environmental monitoring
4. Calibration reports on measuring equipment
5. Test report on potable water
6. Available data on CCP and OPRP monitoring at normal conditions

HAZARD CONTROL PLAN – ORTHODOX BLACK TEA
(Ref: Clause 8.5.4 of ISO 22000:2018)

Critical Control Point Plan Summary									
Plant Name		XYZ Tea Factory							
Critical Control Point (CCP)	**Hazard**	**Critical Limit**	**Monitoring**						
			What	**How**	**Frequency**	**Who**	**Corrective Actions**	**CCP verifications**	**Records**
Drying/ Firing	Survival & proliferation of *Salmonella spp.* *S. aureus* *Escherichia coli*	125 ±2° C	Temperature	Reading oven temperature	Every one hour	Operator	Segregate batch for reprocessing or discard (store airtight) Temperature control adjustments	Oven thermometer calibration	FSMS 04/HC/RC/02 Time & Temperature Records
	Over dried tea	20 minutes	No of minutes	Inlet & outlet air quality	Each batch	Operator	Time control Adjustments	Oven retention time test	Calibration records
	Retention of moisture	3% Moisture	Moisture %	Measure Moisture	Each batch	AFO – Firing	Re-dry to the desired moisture %	Moisture analyser calibration	Visual & Moisture analysis reports.
				Proper maintenance of FSMS 03/WI/FR/03				Internal audit	Internal audit reports

HAZARD CONTROL PLAN – ORTHODOX BLACK TEA

(Ref: Clause 8.5.4 of ISO 22000:2018)

Operational Prerequisite Programme Plan Summary									
Plant Name	**XYZ Tea Factory**								
OPRP	**Hazard**	**Operational limit**	**Monitoring**				**Corrective Actions**	**OPRP verifications**	**Records**
			What	**How**	**Frequency**	**Who**			
Fermentation	Introduction & proliferation of *Salmonella spp.* *Staphylococcus aureus* *Escherichia coli*	2.5 hours 20 ±5° C or 80 ±5° F Hygienic equipment & floor	Fermentation Temperature Hygienic condition of the fermentation room & trolleys/ floor	Temperature & R H control Time control FSMS 03/WI/RR/02 Preventive cleaning ops	Continuous (Every half hour)	Operator AFO – Rolling	Segregate batch in a crate for reprocessing or discard	Thermometer Calibration. Microbiological analysis of intermediate product & floor Internal audit	FSMS 04/HC/RC/01 Time/ temperature/RH record Microbiological analysis reports Calibration records Internal audit reports
Iron removal	Accumulation of Irons through machinery parts of various processing machines & at the field	0% irons at the final product	Use of iron detectors to collect iron at different processing steps	Cleaning Iron detectors after each batch FSMS 03/WI/FR/03 FSMS 03/WI/GR/04	Before start/end of each shift & start of each batch on the go	The operator in relevant machines AFO – Firing/ Sifting / packing	Segregate batch & pass through winnower to recollect iron	Iron detector efficiency calibration Internal audit	FSMS 04/HC/RC/03 Iron collection record Iron detection efficiency reports. Internal audit reports
Packing & Storing	Increase of moisture	5 ±2 % Moisture	Hygienic condition of the sifting room & packing room R H	Measure Moisture Temperature & R H control FSMS 03/WI/PR/05	Each batch	Operator AFO – Firing/ packing	Segregate batch for reprocessing or discard Temperature control	Moisture analyser calibration Internal audit	FSMS 04/HC/RC/04 Moisture analysis reports. Calibration records Internal audit reports

UPDATING OF INFORMATION SPECIFYING PRPS AND HAZARD CONTROL PLAN

(Ref: Clause 8.6 of ISO 22000:2018)

Scope

The scope covers all preliminary information of hazard analysis, hazard control plan, PRPs and OPRPs needed to maintain food safety management system effectively and up to date.

Activities

1. The FST shall update the system information based on input from communication, external as well as internal (clause 7.4), input from other information concerning the suitability, adequacy, and effectiveness of the FSMS, output from the analysis of results of verification activities (clause 8.8) and output from management review (clause 9.3).

2. System updating activities shall be retained as documented information and reported as input to the management review (clause 9.3).

3. Following documents shall be amended and reviewed according to information communicated internally and externally to keep them up to date.
    1. Raw and packing materials characteristics
    2. Characteristics of end products
    3. Intended use
    4. Flow diagrams
    5. Processing steps
    6. Hazard control plan
    7. PRPs
    8. Control measures

References

1. Description of raw and packing material
2. Description of finished products
3. Intended use
4. Process flow diagrams
5. Descriptions on control measures
6. PRPs
7. OPRPs
8. CCPs
9. Hazard control plan

CONTROL OF MONITORING & MEASURING

(Ref: Clause 8.7 of ISO 22000:2018)

Scope

The scope covers all monitoring and measuring equipment used in operations that have an impact on food safety.

Activities

1. M/s XYZ Ltd. has prepared a documented procedure that specifies monitoring and measuring methods and equipment in use, which are adequate for the monitoring and measuring activities related to the PRP(s) and the hazard control plan.
2. The equipment used for monitoring and measuring activities shall be selected by determining the measurement to be made and the accuracy required. Each department shall maintain a list of monitoring and measuring equipment.
3. The monitoring and measuring equipment shall be calibrated and verified against measurement standards traceable to international and national measurement standards periodically. Where no standards exist, the basis used for calibration or verification shall be retained as documented information.
4. If equipment is found to be out of calibration, the measurement made with the equipment shall be rechecked for accuracy and adjusted or re-adjusted as necessary.
5. The measuring equipment shall be stored and maintained in such a way that it is protected from damages or deterioration and adjustment that would invalidate the measurement results.
6. The results of calibration and verification shall be retained as documented information.

Reference

1. Control of monitoring and measurement of equipment and methods – PRP 04

VERIFICATION PLANNING

(Ref: Clause 8.8 of ISO 22000:2018)

Scope

The scope covers all planned arrangements of evaluation of individual verification results and analysis of results of verification activities in the food safety management system.

Activities

1. M/s XYZ Ltd. shall establish, implement and maintain verification activities that define the purpose, methods, frequencies, and responsibilities for the verification activities.
2. The FST has defined each PRP, OPRP, and the CCP with their intended results and the method of verification such a way that, when there is a deviation from the expected results that can be identified before it will affect the food safety of orthodox black tea manufactured.
3. The FST shall evaluate verification activities to conform to the implementation. If any deviations are found, appropriate actions shall be taken to achieve the required conformity.
4. The individual verification activities shall confirm the PRPs are implemented and effective, input to the hazard analysis is updated periodically, the hazard control plan is implemented and effective, hazard levels are within identified acceptable levels, and other actions determined by the FST are implemented and effective.
5. The FSTL shall ensure that verification activities are not carried out by the person responsible for monitoring the activity or the control measures.
6. The FST shall analyse the results of verification activities, including the results of the internal audit and external audits. If any deviations are found, appropriate actions shall be taken to achieve the required conformity.
7. The results of the analysis shall be used for improvements of FSMS, to identify the trends which may lead to potentially unsafe products, to plan the internal audit, and to evaluate the effectiveness of the correction and corrective actions.
8. The above activities shall be recorded and reported to the food safety team and management review meetings.

Reference

1. Hazard control plan
2. Hazard control records
3. PRP records
4. System validation records
5. Validation guide

CONTROL OF PRODUCT AND PROCESS NONCONFORMITIES

(Ref: Clause 8.9 of ISO 22000:2018)

Scope

The scope covers the aspects of corrections, corrective actions done to the processes, handling of unsafe products, and withdrawals of non-conforming products.

Activities

1. M/s XYZ Ltd. has appointed designated persons with sufficient competence and authority to evaluate the data derived from the monitoring of OPRPs and at CCPs and to initiate corrections and corrective actions.
2. When applicable, verification is based on testing of end product samples or direct process samples, which shows nonconformity with the acceptable level of the food safety hazard (clause 8.5.2.2), the affected lots of product shall be handled as potentially unsafe (clause 8.9.4).

Corrections

1. The data derived from the OPRPs and CCPs shall be evaluated by the Factory manager, relevant Assistant Factory Officer, Administrative Officer, and General Manager to ensure that when critical limits at CCPs and/or action criteria for OPRPs are not met, the products affected are identified and controlled with regard to their use, release and initiate corrective actions.
2. All corrective actions taken shall be recorded.
3. The products shall be declared potentially unsafe when the critical limits are not met, and potentially unsafe products shall be segregated and handled according to the procedure for handling of unsafe products.
4. FST shall retain documented information to describe corrections taken on nonconforming products and processes, including the nature of the nonconformity, the root causes of the nonconformity, the consequences as a result of the nonconformity, and the traceability information related to the lots of nonconforming products.

Corrective actions

1. The FST shall apply corrective actions as per clause 8.9.3 when critical limits at CCPs and/or action criteria for OPRPs are not met.
2. Immediate corrections shall be taken to prevent operations being moving out of the control limits, which will affect food safety.
3. In the case of produced black tea is unsafe for human consumption, it will be evaluated with respect to the cause and removed from the premises.
4. M/s XYZ Ltd. has an established documented procedure to identify and assess and handle the nonconformity, which specifies appropriate actions to identify and eliminate the cause of detected nonconformities, to prevent a recurrence, and to return the process to control after a nonconformity is identified.

Handling of potentially unsafe products

1. The FST shall take actions to prevent potentially unsafe products from entering the food chain unless it is possible to demonstrate that the food safety hazards of concern are reduced to the defined acceptable levels or will be reduced to

identified acceptable levels prior to entering the food chain and the product still meets the defined acceptable levels despite the nonconformity.

2. Products identified as potentially unsafe shall be held under the control of the General Manager until they have been evaluated. If products that have left the control of the organization are subsequently determined to be unsafe, the organization shall notify relevant interested parties and initiate a withdrawal/recall (clause 8.9.5).
3. The controls and related responses from relevant interested parties and authorization for dealing with potentially unsafe products shall be retained as documented information.
4. Each lot of products affected by the nonconformity shall be evaluated.
5. Products affected by failure to meet critical limits at CCPs shall not be released but be handled in accordance with clause 8.9.4.3.
6. Products affected by failure to meet action criteria for OPRPs shall only be released as safe when evidence other than the monitoring system demonstrates that the control measures have been effective or the combined effect of the control measures for that particular product conforms to the performance intended.

Withdrawals

1. If a batch of products is identified as unsafe, a decision to withdraw shall be taken by a team headed by General Manager.
2. The affected batch shall be withdrawn from the market as soon as possible according to the procedure for withdrawals.
3. The complete and timely withdrawal/recall of end products that have been identified as potentially unsafe will be carried out by appointing competent persons having the authority to initiate and carry out the withdrawal/recall and establishing and maintaining documented information for notifying relevant interested parties (e.g., statutory/regulatory authorities, customers and/or consumers), handling withdrawn/recalled products as well as products still in stock and performing the sequence of actions to be taken.
4. Withdrawn/recalled products and end products still in stock shall be secured or held under the control of the organization until they are managed in accordance with 8.9.4.3.
5. The cause, extent, and result of a withdrawal/recall shall be retained as documented information and reported to the top management as input for the management review (clause 9.3).
6. The FST shall verify the implementation and effectiveness of withdrawals/recalls through the use of appropriate techniques (e.g., mock withdrawal/recall or practice withdrawal/recall) and retain documented information.

References

1. Food safety procedure for correction and corrective action – FSP 06
2. Food safety procedure for handling of potentially unsafe products – FSP 07
3. Food safety procedure for withdrawals/recalls – FSP 08

PERFORMANCE EVALUATION

(Ref: Clause 9.1 of ISO 22000:2018)

Scope

The process covers all monitoring, measurement, analysis, and evaluation conducted for the food safety management system.

Activities

1. M/s XYZ Ltd. has planned what needs to be monitored and measured with appropriate methods for monitoring, measurement, analysis, and evaluation to ensure valid results while specifying when and where monitoring and measuring shall be performed with the timeline and specific persons for analysis and evaluation of results from monitoring and measurement.
2. The FST shall maintain and retain relevant information as evidence of the results, which shall be used for the evaluation of performance and the effectiveness of the FSMS.
3. The FST shall analyse and evaluate appropriate data and information arising from monitoring and measurements, including the results of verification activities related to PRPs and the hazard control plan (clause 8.8), the internal audits (clause 9.2), and external audits.
4. The analysis shall be carried out using statistical methods, root cause analysis, graphs, charts, or as reports with logical sequence based on scientific evaluation methods.
5. The analysis shall be carried out to confirm if the overall performance of the system meets the planned arrangements and the FSMS requirements established or to identify the need for updating or improving the FSMS.
6. The analysis shall be used to identify trends that indicate a higher incidence of potentially unsafe products or process failures and establish information for planning of the internal audits related to the status and importance of areas to be audited, and to provide evidence that any corrections and corrective actions that have been taken are effective.
7. The results of the analysis and any resulting activities shall be retained as documented information by the FST and shall be reported to top management and used as input to the management review (clause 9.3) and for the updating of FSMS (clause 10.2).

Reference

1. Annual audit plan with audit reports
2. Documented results, measurement, and analysis
3. System updating activities planned and completed
4. Continual improvements carried out for the system

INTERNAL AUDIT
(Ref: Clause 9.2 of ISO 22000:2018)

Scope

This process covers all internal audits conducted for the food safety management system.

Activities

1. M/s XYZ Ltd. has prepared and implemented an internal audit procedure that includes the relevant instructions and methods of conducting, evaluating, and documenting the audit results.
2. M/s XYZ Ltd. shall conduct internal audits every three (3) months to provide information on whether the FSMS conforms to the set requirements of the food safety management system and the requirements of ISO 22000:2018, which is effectively implemented and maintained.
3. Annual audit plan shall be prepared by Consultant - FSMS and circulate to relevant personnel. An audit schedule shall be issued one week prior to the audit date.
4. The importance of the process, changes that occurred or planned, and results of the previous audit shall be taken into consideration for planning of audits.
5. It shall indicate the division/department to be audited, audit scope, and members of the audit team.
6. Nonconformity reports shall be raised if any nonconformity is found. The team leader shall write an audit report, and follow-up action regarding the corrective actions shall be carried out.
7. Follow-up activities shall include the verification of the actions taken and the reporting of the verification results.
8. The results of the previous audits, follow up and changes to the system based on audit results shall be maintained as documented information.

Reference

5. Annual audit plan
6. Internal food safety audit report
7. Internal food safety audit summary
8. Food safety procedure for internal food safety audits – FSP 09

MANAGEMENT REVIEW
(Ref: Clause 9.3 of ISO 22000:2018)

Scope

The scope covers the meetings conducted by the top management to review the suitability, adequacy, and effectiveness of the food safety management system.

Activities

1. M/s XYZ Ltd. has prepared and implemented a procedure to conduct management review meetings that explain the process and activities.
2. All management reviews shall be conducted with the participation of the food safety team and management team.
3. A food safety management review meeting shall be held once every three months, where top management shall review the organization's FSMS to ensure its continuing suitability, adequacy, and effectiveness.
4. The schedule shall be noticed one week in advance, and the minutes shall be circulated by the food safety team leader among the members.
5. The food safety team leader shall retain documented information as evidence of the results of the management review.

Reference

1. Food safety procedure for management reviews – FSP 10
2. Minutes of management review meetings

IMPROVEMENT

(Ref: Clause 10.1. 10.2 & 10.3 of ISO 22000:2018)

Scope

This process covers all corrective actions and the continuous improvement carried out for the food safety management system.

Activities

Nonconformities and corrective action

1. M/s XYZ Ltd. has empowered the FST to react to the nonconformities as applicable by taking action to control and correct them while dealing with the consequences.
2. The FST shall evaluate the need for action to eliminate the cause(s) of the nonconformity, in order that it does not recur or occur elsewhere, in the following manner:
    a. review the nonconformity;
    b. determine the causes of the nonconformity;
    c. determine if similar nonconformities exist or could potentially occur;
3. The FST shall implement any action needed as to the nonconformity and review the effectiveness of any corrective action taken while making changes to the FSMS, if necessary, where the corrective actions shall be appropriate to the effects of the nonconformities encountered.
4. The FST shall retain documented information as evidence of the nature of the nonconformities and any subsequent actions taken with the results of any corrective action.

Continual improvement

1. M/s XYZ Ltd. shall continually improve the suitability, adequacy, and effectiveness of the FSMS to enhance the operations.
2. Corrective action reports, preventive action reports, nonconforming product reports, and internal audit reports shall be analyzed every three months to see the effectiveness and improvement of FSMS.
3. Top management shall ensure that the organization continually improves the effectiveness of the FSMS through the use of communication (clause 7.4), management review (clause 9.3), internal audit (clause 9.2), analysis of results of verification activities (clause 8.8.2), validation of control measures and combinations of control measures (clause 8.5.3), corrective actions (clause 8.9.3) and FSMS updating (clause 10.3).
4. Management review, internal audits, evaluation of individual verification results, analysis of results of verification activities, validation of control measure combinations, corrective actions, FSMS updating, and communication shall be considered for the effectiveness of the food safety management system.

Updating the food safety management system

5. M/s XYZ Ltd. has authorized FST to evaluate the FSMS at planned intervals where FST shall review the FSMS annually, and the team shall then consider whether it is necessary to review the hazard analysis (clause 8.5.2), the established hazard control plan (clause 8.5.4) and the established PRPs (clause 8.2).

6. The top management shall review the suitability of its FSMS through management review and ensure it is continually updated.
7. The FST shall update FSMS based on input from communication that is external as well as internal (clause 7.4) and input from other information concerning the suitability, adequacy, and effectiveness of the FSMS as well as the output from the analysis and evaluation of monitoring and measurement (clause 9.1); and output from management review (clause 9.3).
8. FST shall ensure that the FSMS is continually updated and retain system updating activities as documented information and reported as input to the management review (clause 9.3).

Reference

1. Internal audit results
2. Corrective action reports
3. Preventive action reports
4. Nonconforming product reports

ISO 22000 FOOD SAFETY PROCEDURE

FSP 01 - CONTROL OF DOCUMENTED INFORMATION

Distribution:

1. Director/General Manager
2. Factory Manager
3. Accountant
4. Factory Officer
5. Quality Officer
6. Administration Officer
7. Assistant Factory Officer – Withering
8. Assistant Factory Officer – Rolling
9. Assistant Factory Officer – Firing
10. Assistant Factory Officer – Sifting
11. Consultant – FSMS

1.0 Objective

To ensure the documented information related to the food safety management system are approved, reviewed, and updated as necessary, properly identified and available where and when needed, and legible prior to issue. The procedure also facilitates their use, the authority of viewing, amendments, and prevents the unintended use of obsolete documents.

2.0 Scope

The scope covers all the documented information of internal and external origin, including the amendments and revisions.

3.0 Applicable documents

Food Safety Manual	-	Level 01 Documents
Food Safety Procedures	-	Level 02 Documents
Prerequisite Programmes	-	Level 02 Documents
Work Instruction, attachments	-	Level 03 Documents
Check lists, forms, flow charts, books	-	Level 04 Documents
reports, records, guides, and registers		
internal and external communication		

4.0 Responsibilities

Overall	– General Manager
Direct	– Factory Manager/Food Safety Team Leader
	– Food Safety Team
	– All Department/Section Heads

4.0 Activities

4.1 Document approval

The document prepared shall be reviewed and approved by relevant personal authorized to do so at the bottom of every document prior to issue.

The food safety manual shall be approved by General Manager, and the circulation shall be controlled by Food Safety Team Leader.

4.2 Review and update

Any change to a document/documented information shall be proposed through a memorandum for document and data change (FSMS 04/FSP/01/F/02) to change document and forward to Factory Manager. The Factory Manager shall prepare the specified document with the person who prepared the original document. The document shall be reviewed and approved by authorized personal prior to circulating. All obsolete documented information shall be returned by respective personal to the Factory Manager. If the obsolete documented information is required to keep, place the stamp "OBSOLETE BUT KEEP FOR SPECIAL PURPOSE" otherwise destroy.

4.3 Revision status of the documents

Revision status, date issued, and the document number shall be stated at the bottom of each page of the document.

4.4 Amendments and revision

4.4.1 **Minor changes** – which will not affect to the content shall be done by the person who prepared the original document and sends for review and approval by the person who reviewed and approved the original document (FSMS 04/FSP/01/F/04). All amendments shall be recorded in the amendment and revision record sheet.

4.4.2 A change that will affect the content of a clause, FSP, or process shall be treated as a revision. The revision status has to be changed. The document shall be review and approve by the person who reviewed and approved the original document.

4.5 Document circulation

Copies of the approved document shall be circulated to relevant personal together with a document acknowledgement memo (FSMS 04/FSP/01/F/01). Each page of all documents shall be stamped with "Controlled Copy" to ensure that relevant versions of the applicable documents are available at relevant places. A list of master copies and controlled copies shall be maintained. (FSMS 04/FSP/01/F/03)

4.6 All documents of external origin (standard, regulations, etc.) shall be stamped with "APPROVED" by General Manager and circulated to relevant points. A list of documents of external origin shall be maintained (FSMS 04/FSP 01/F/05). Validity of the documents of external origin shall be checked annually by Consultant - FSMS.

4.7 Document numbering system.

4.7.1 Food Safety Manual – Level 01 (E.g., FSMS 01/03)

Copies of the Food Safety Manual shall be numbered as 1, 2, 3, etc.

4.7.2 Food Safety Procedures – Level 02 (E.g., FSMS 02/FSP/01)

Food safety Procedures shall be numbered as FSP/01, FSP/02, FSP/03, etc.

Prerequisite Programmes shall be numbered as PRP/01, PRP/02, PRP/03, etc.

4.7.3 Other documents – Level 03/Level 04

- Work Instructions shall be numbered as WI/Abbreviation of place applicable/ serial number (E.g., D. Ref: FSMS 03/WI/ RR/02)
- Records or evidence document shall be numbered as PRP or FSP/type of document /01 to infinite numbers (E.g., D. Ref: FSMS 04/PRP/03/CL/01)
- Safety plans shall be numbered as SP(serial number)/Abbreviation of place applicable/01
- The format shall be numbered as F(serial FSP number/ number)
- Reports shall be numbered as RP (FSP number/serial number)
- Records shall be numbered as R (FSP number/serial number)
- If it is not described in an FSP, those formats shall be numbered as F(serial number)/Abbreviation of department/(WA, RR, FR, or GR) (If necessary to specify)

4.8 Abbreviations of divisions/ departments used in Quality system documents

WA	-	Withering Area
RR	-	Rolling Room
FR	-	Firing Room
GR	-	Grading Room
STORE	-	Stores
TPT	-	Transport

5.0 Records

5.1 Identification

Food Safety Records are those which describe the result of some activities of food safety management system such as inspection, audits, customer complaints, etc., which includes all the records generated from the FSMS, including the electronic media records.

5.2 Storage and Protection

All records shall be stored to prevent damage or deterioration. All the records shall be protected from displacement by indexing.

All electronic media records shall be password protected when applicable. Access to the different functions of such a system depends on the requirement and has to be authorized by General Manager.

5.3 Retrieval

All records shall be filed, labelled, and stored systematically so that quick retrieval is possible.

4.4 Retention

All records shall be kept in good condition by the person responsible until the retention time is over. The retention times for different records are given in Annexure 1.

4.5 Disposition

Records shall be disposed of when the retention time is over by the department manager /system administrator.

5.0 References

5.1	FSMS 04/FSP/01/F/01	- Document Acknowledgement Memo
5.2	FSMS 04/FSP/01/F/02	- Memorandum for Document and Data Changes
5.3	FSMS 04/FSP/01/F/03	- List of Master Copies or Controlled Copies of Internal Documents
5.4	FSMS 04/FSP/01/F/04	- Amendment/Revision Record Sheet
5.5	FSMS 04/FSP/01/F/05	- List of Master Copies or Controlled Copies of External Documents
5.6	FSMS 04/FSP/01/F/01	- List of Food Safety Records

D. Ref: FSMS 04/FSP/01/F/01

XYZ TEA FACTORY

DOCUMENT ACKNOWLEDGEMENT MEMO

Distribution to:

Document reference no:

No of page:

I herewith attach a document(s), which is (or are) new/a revision(s)/having amendment(s). Ensure to use the attached document/s for all relevant future activities.

Please acknowledge the receipt of the document/s by filling the latter portion and returning this memo together with the obsolete document/s within four (4)working days.

....

Date Signature FM

To: Factory Manager

This is to acknowledge the receipt of the .... I herewith attach/ not attach .... the obsolete documents.

Date:........................ Designation:...............

Signature:....................

D. Ref: FSMS 04/FSP/01/F/02

XYZ TEA FACTORY

MEMORANDUM FOR DOCUMENT AND DATA CHANGES

Type of change proposed: revision/amendment

Original document/Data title	....	No	....
Revision no	....	Location	....
Document holder	....	Reviewed by	....
Approved by	....	Date issued	....

To: General Manager
From: .... (though) Department/ section head

I propose the above change as per the attached draft for the following reasons ....
....
Please prepare in the required format and forward for review, approval, and implementation if suitable.

Signature: .... Date: ....

To: Factory Manager
From: Approving authority.....................................
Revision number-new (increase the old number by 1).......... Date issued at origin ……….......... Date issued new.................................

Approved. Please distribute and implement

Signature: .... Date: ....

Document or date finalized in proper format, review, and approval completed. Distribution completed and recalled all obsolete items. The revision was added to the master copy, the obsolete master copy withdrawn, and the master list amended.

Factory Manager

Signature: .... Date: ....

D. Ref: FSMS 04/FSP/01/F/03

XYZ TEA FACTORY

FORMAT FOR LIST OF MASTER COPIES OR CONTROLLED COPIES OF INTERNAL DOCUMENTS

Document Title	Document Code	Date Issued	Revision Number	Location	Document Holder/s

D. Ref: FSMS 04/FSP/01/F/04

XYZ TEA FACTORY

AMENDMENT/REVISION RECORD SHEET

Food Safety Procedure - FSP (Includes attachments and work instructions referred in the FSP)

Document Title	Document Code	Date Issued	Revision Number	Location	Document Holder/s	WI/ Attachment

D. Ref: FSMS 04/FSP/01/F/05

XYZ TEA FACTORY

FORMAT FOR LIST OF MASTER COPIES OR CONTROLLED COPIES OF

EXTERNAL DOCUMENTS

Document Title	Effective Date	Revision No.	Location	Document Holder/s

D. Ref: FSMS 04/FSP/01/F/06

XYZ TEA FACTORY

LIST OF FOOD SAFETY RECORDS

Food Safety Record Code	Name of Records	Record Holders	Retention Time (Years)

ISO 22000 FOOD SAFETY PROCEDURE

FSP 02 – STATUTORY, REGULATORY, AND CUSTOMER REQUIREMENTS

Distribution:

1. Director/General Manager
2. Factory Manager
3. Accountant
4. Factory Officer
5. Quality Officer
6. Administration Officer
7. Assistant Factory Officer – Withering
8. Assistant Factory Officer – Rolling
9. Assistant Factory Officer – Firing
10. Assistant Factory Officer – Sifting
11. Consultant – FSMS

1.0 Objective

To make all relevant personnel informed duly and timely on statutory, regulatory, and customer requirements related to the food safety of the orthodox black tea manufacturing process and how to address them through the food safety management system.

2.0 Scope

This procedure covers relevant interested parties, their requirements on food safety concerns, and the application of appropriate standard methods to address them to achieve set compliances.

3.0 Responsibilities

Overall	– General Manager
Direct	– Food Safety Team Leader (Internal/External Liaison)
	– Factory Manager
	– All the employees of the company

4.0 Activities

4.1 Interested Parties

4.1.1 The management considers stakeholders, persons, or organizations that can influence our food safety/food quality, business continuity, and persons or organizations that can be affected by our food safety/food quality or business continuity activities. Thus, interested parties could include the following, but not limited to;

4.1.2 **Customers:**

The customer includes auctioneer, immediate buyers, intermediators, brokers, packers, shippers, value-added producers, wholesalers, and retailers in local or international markets, who are directly or indirectly related and communicate along the supply channel with specific food safety and other product-related requirements.

M/s XYZ Ltd. shall attract and retain customers by offering products, services that meet their food safety requirements along with excellent customer service.

4.1.3 **Employees:**

The direct and indirect employees working and paid by the M/s XYZ Ltd., as compensation for the service acquired irrelevant to the position the activities carried out will consider for their active contribution towards food safety and quality of orthodox black tea manufactured.

The company makes sure the availability of people with the motivation to remain as contributing members of the organization and develop the skills necessary to provide a competitive edge.

4.1.4 **Suppliers:**

Green leaf suppliers and other non-raw material suppliers are critical to our activities. Thus, it is very important to maintain supplier compliances and close supplier relationships, which are effective ways to remain competitive and secure the resources required.

Hence, supplier compliances will be considered as the major requirement while providing resources required for them to implement relevant compliances set by the FSMS and specific interested parties.

4.1.5 **Investors:**

The company is considering the views of its investors, bankers, and other financiers as their views are critical factors to the success of the implementation of relevant standards and legislative requirements to sustain their investments.

4.1.6 **Competitors:**

The FST shall identify and evaluate competitors in relation to the food safety of their similar product offerings and document the outcome.

4.1.7 **Others:**

The following groups are also important, but the considerations are given only for issues arising due to the food safety or hazard occurrences: Shareholders, Owners, Management, Employees, Trade unions, Partners, Client, Government agencies, Society, any other person or organizations interested in the organization's activities.

4.2 The applicable food legislation (in the country of sale or use including maximum residue limits, food safety concerns, packaging, product description, net weights, nutritional information, allergen, and additive or identity-preserved foods labelling and any other requirements under given food legislation and the established industry codes of practices) according to buyer profiles/auction data shall be collected and updated as appropriate based on importing country requirements.

4.3 The statutory and regulatory requirements may include, but not limited to ISO 22000:2018, ISO 22002-2 PRP requirements, CODEX GAP, GHP, GMP Principles, ISO 9001:2015, ISO 3720:2011, 5S, SLS 135, Sri Lanka Tea Board Circular No.: AL/MQS-Rev/2010 26th April 2010 by the SLTB with Annexure - A for the Sri Lankan Origin teas.

4.4 Collect information on supplier compliances, including pesticide/fertilizer applications, any unintended biological, physical, or chemical contaminations, and identified food safety hazards through the supply chain, communicated by customers, or available in the public domain for the analysis and compliances requirements.

4.5 Collect information related to food safety incidents related to tea growing, harvesting, transportation, manufacturing, packaging, and delivery to the auction. Collect information related to food safety incidents related to tea growing, harvesting, transportation, manufacturing, packaging, and delivery to the auction.

4.6 Obtain specific customer requirements, if any, communicated by the buyers through the auctioneer.

4.7 Consider public perceptions, social media, mass media as well as employee-raised issues relevant to the food safety of the tea supply chain.

4.8 Incidents, references, specific food safety requirements, standards, guidelines, and compliances set by the exporting countries or specific buyers.

4.9 Consider all the internal and external communications related to food safety or product issues recorded in the documented information.

4.10 M/s XYZ Ltd. shall plan its actions to address identified risks and opportunities while integrating and implementing the actions into the FSMS processes, which can evaluate the effectiveness of those actions.

4.11 The risk and the opportunities shall be addressed to assure the FSMS can achieve its intended outcome(s)/results while preventing or reducing undesired effects and achieve continual improvement.

4.12 The FSMS shall be planned to comply with ISO 22000, including the aforementioned statutory/regulatory or customer requirements, and to address risk and the opportunities generated through the manufacturing process.

4.13 Consider all the given requirements and synchronize those set requirements with ISO 22000 requirements while applying the PDCA cycle and the HACCP principles.

4.14 Conduct the hazard analysis and hazard assessment according to set requirements or, if necessary, go beyond what ISO 22000 has recommended based on the collected information (i.e., a customer request for FSSC 22000, SQF, or BRC for a specific country). If there is no necessity to go beyond the general requirements set by the ISO 22000, then follow the aforementioned requirements to achieve complete reliability for the food safety compliance of the generic tea production system as a norm.

4.15 All operations within the scope defined shall be planned, executed, validated, and verified to reduce or eliminate hazards that are possible to occur in the products.

4.16 Any specific customer requirements provided, communicated, or publically available shall be considered case by case, and relevant compliances will be covered accordingly.

4.17 The FSMS shall maintain, update, and verify as appropriate and retain documented information on the requirements that are identified, communicated, provided, or regulated.

4.18 M/s XYZ Ltd. shall plan, execute, and document periodical audits, management reviews, and updating activities to monitor and verify the system compliances and appropriately communicated to relevant interested parties.

4.19 Any change, customer compliance, statutory/regulatory changes, and updates that can have an impact on food safety shall be communicated to the food safety team.

4.20 Additional information regarding the planning and execution process shall be given as appropriate in the relevant food safety procedures, prerequisite programs, work instructions, and guides provided.

4.21 The FSMS planning shall also explain the additional information related.

4.22 The food safety manual is designed to address generic requirements or most of the minor matters that are not required a documented procedure to explain the execution, as well as a linking document to the entire documentation system.

5.0 References

5.1 The food safety procedures

5.2 The prerequisite programs

5.3 Hazard Assessment

5.4 Work instructions

ISO 22000 FOOD SAFETY PROCEDURE

FSP 03 - COMMUNICATION

Distribution:

1. Director/General Manager
2. Factory Manager
3. Accountant
4. Factory Officer
5. Quality Officer
6. Administration Officer
7. Assistant Factory Officer - Withering
8. Assistant Factory Officer - Rolling
9. Assistant Factory Officer - Firing
10. Assistant Factory Officer - Sifting
11. Consultant - FSMS

1.0 Objective

To make all relevant personnel informed duly and timely on all food safety related issues and to make timely decisions on issues related to food safety.

2.0 Scope

This procedure covers all communications of the organization (internal and external) related to food safety.

3.0 Responsibilities

Overall	- General Manager
Direct	- Food Safety Team Leader (Internal/External Liaison)
	- Factory Manager

4.0 Activities

4.1 External Communication

4.1.1 Any food safety related information (instructions, guidelines, and requirements) shall be communicated timely to interested parties in written format. The Factory Manager shall communicate with suppliers and contractors regarding materials and services. The General Manager shall communicate with contractor/service providers for the work environment, infrastructure, and equipment. The food safety team leader shall communicate with any interested parties related to food safety issues, compliances, and operational matters relevant to FSMS.

4.1.2 The supplier compliances, including pesticide/fertilizer applications, any unintended physical or chemical contaminations (other leaf particles, insect parts, or droppings), and identified food safety hazards through the supply chain shall be communicated to green leaf suppliers by Food Safety Team Leader thorough supplier awareness biannually.

4.1.3 Product information, including intended use, safe handling, specific storage requirements, shelf life, and any other relevant information shall be communicated in a due manner. This information may be included in product labels or contracts where appropriate. Approval of the General Manager shall be obtained before issuing product safety related communication.

4.1.4 The communication related to food safety with distributors shall be done through the Factory Manager or General Manager to auctioneers. Communication regarding recall/ withdrawals shall also be done through the above channel. The complaints shall be directly communicated to the Factory Manager and General Manager.

4.1.5 Statutory and regulatory authorities and any other interested parties that may have an interest or impact on food safety shall be informed as appropriate on issues concerning them. Such communication shall be done by the food safety team leader or under the approval of the General Manager.

4.1.6 Any inquiries, customer feedback, or any other information from the food chain, including customers, suppliers, and statutory and regulatory authorities, shall be communicated to Food Safety Team Leader. Customer complaints and related activities also shall be communicated to the Factory Manager and General Manager. All above information and issues shall be documented and communicated to the food safety team.

4.1.7 All food safety related external communications shall be copied to Food Safety Team Leader, and, thereby, the FST shall be acknowledged.

4.2 Internal Communication

4.2.1 Any information from either customer, regulatory authorities, or any other interested parties that have an impact in FSMS shall be communicated to relevant Assistant Factory Officers, supervisors, or workforce, effectively by means of e-mail or printed notice with a copy to Food Safety Team Leader.

4.2.2 Internally generated information having an impact on food safety shall be communicated to relevant Assistant Factory Officers, supervisors, and workforce duly via e-mail or as printed information and records shall be maintained, in electronic or printed format.

4.2.3 Any food safety related information, including violation of CCPs, violation of OPRPs, and PRPs, shall be taken into the notice of relevant Assistant Factory Officers by any individual who notices it. This information shall be recorded, and action taken may also be recorded accordingly.

4.2.4 The food safety team shall be informed by relevant Officers regarding the following changes which may have an impact on FSMS.

- Products or new products
- Raw materials, ingredients, and services
- Production systems and equipment
- Production premises/location of equipment/surrounding environment
- Cleaning and sanitation programmes
- Packaging storage and distribution systems
- Personnel qualification levels and/or allocation of responsibilities and authorisations
- Statutory and regulatory requirements
- Knowledge regarding food safety hazards and control measures
- Customer, sector, and other requirements that the organisation observes
- Relevant inquiries and communications from external interested parties
- Complaints, risks, and alerts indicating food safety hazard associated with the end product
- Other conditions that have an impact on food safety
- Resource needs, improvement, and the effectiveness of the FSMS.
- Revision of the organization's food safety policy and related objectives.

4.2.5 The FST shall collect, include and utilize that information during the implementation or while updating the FSMS and also include them into the management review as inputs.

5.0 References

5.1 Company database on external and internal communication

6.0 Records

6.1 Internal communication records

6.2 External communication records

ISO 22000 FOOD SAFETY PROCEDURE

FSP 04 – PRODUCT IDENTIFICATION AND TRACEABILITY

Distribution:

1. Director/General Manager
2. Factory Manager
3. Factory Officer
4. Quality Officer
5. Administration Officer
6. Assistant Factory Officer – Withering
7. Assistant Factory Officer – Rolling
8. Assistant Factory Officer – Firing
9. Assistant Factory Officer – Sifting
10. Auctioneer
11. Consultant – FSMS

1.0 Objective

To ensure information and products are related along the supply chain and to establish and apply a traceability system that enables the identification of end product lots in relation to their input, processing and delivery records, and timely use in the event of a product withdrawal.

2.0 Scope

The scope permits linkage one step back to immediate suppliers and one step forward in the distribution of the end product. The system shall meet all statutory and regulatory requirements and provide for record-keeping while covering the identification of raw and packaging materials, in-process material, and final products.

3.0 Responsibilities

Overall	- General Manager
Direct	- Food Safety Team Leader
	- All Assistant Factory Officers
	- Auctioneer

4.0 Activities

4.1 The supplier's Register number and the quantity (kg) of raw materials are entered into green leaf receiving register with supply line (area) identification code/names.

4.2 Withering troughs shall be tagged according to supply line name or identification code.

4.3 Withered green leaves shall be released according to the systematic 250Kg - 300Kg batches as charging required by the rolling room.

4.4 Dhools shall be coded and fermented according to pre-identified batch codes where time chart and dhool release time shall be recorded with a consecutive numbering system containing date, time, and the type of dhool with batch number.

4.5 The firing room shall record the time, type of dhool, and storage bin of the intermediate product while releasing to sifting.

4.6 The sifting area shall be equipped with numbered creates to handle specific grades while recording the specific grades and batches.

4.7 The processing records shall include the batch ID-Code including the time/date at which the process was carried out, in addition to other information relating to the processing conditions that shall be required in terms of product safety or quality assurance.

4.8 Intermediate products shall be segregated according to the type of variety to be manufactured (Eg. such as leafy, pekoe, FBOP, dust, fannings, and fibre into separate demarcated areas inside the grading room with the batch card for each batch) while using FSMS 04/FSP/04/BC/01, FSMS 04/FSP/04/BC/02 and FSMS 04/FSP/04/BC/03 (BC - Blend Card) for traceability. Colour codes shall be used to segregate product dhools at the drying room as well as the next specific machine operation of the grading room.

4.9 When the dried tea issue to production, the batch number shall be entered in Goods Movement Report when those materials use for blending/mixing the batch no shall be entered in batch process records.

4.10 The batch code of the final product consists of the date of production, variety, and time. This batch number is recorded in the batch process record.

4.11 Details of other materials and packing materials shall be obtained from the suppliers with the supplier quality assurance program.

4.12 Individual batches shall be identified by product coding throughout the time that they are within the factory, and batch separation shall be maintained throughout the processing operation and storage while linking batch codes to production records.

5.0 References

5.1 Food safety procedure for withdrawals/recalls – FSMS 02/FSP/08

6.0 Records

6.1 Green leaf receiving register – FSMS 04/FSP 04/RC/01

6.2 Internal batch moving records

Withering	– FSMS 04/PRP/05/RC/03
Rolling	– FSMS 04/HACCP/RC/01
Firing	– FSMS 04/HACCP/RC/02
Sifting	– FSMS 04/FSP/04/BC/01
	– FSMS 04/FSP/04/BC/02
	– FSMS 04/FSP/04/BC/03
Packing	– FSMS 04/HACCP/RC/04

6.2 Product transport record to auctioneer stores

Intake Ticket		
		D. Ref: FSMS 04/FSP/04/BC/01
XYZ Tea Factory		
Code	**Source**	**Serial No**
Quantity (Kg)		**Time**
Sub Batch		**Date**

Batch Card			
			D. Ref: FSMS 04/FSP/04/BC/02
XYZ Tea Factory			
Date	**Source Code**	**Grade**	**Quantity (Kg)**
Batch Code	**Residue Batch**	**Grader**	

Blending Ticket		
		D. Ref: FSMS 04/FSP/04/BC/03
XYZ Tea Factory		
Code	**Source Batch Codes**	**Date**
Quantity (Kg)		**Time**
Grade		**Final Product**
Comments		

ISO 22000 FOOD SAFETY PROCEDURE

FSP 05 - EMERGENCY PREPAREDNESS AND RESPONSE

Distribution:

1. Director/General Manager
2. Factory Manager
3. Factory Officer
4. Quality Officer
5. Administration Officer
6. Assistant Factory Officer - Withering
7. Assistant Factory Officer - Rolling
8. Assistant Factory Officer - Firing
9. Assistant Factory Officer - Sifting

1.0 Objective

To prevent contamination or deterioration of products, raw material, or intermediate products while controlling any unsafe products during and after an emergency situation.

2.0 Scope

This Procedure covers the action to be taken in an emergency situation within the scope of the food safety management system to prevent contamination of products. These events shall include incidents such as fire, flooding, bio-terrorism and sabotage, energy failure, vehicle accidents, contamination of the environment, various types of weather-related events, or the impact of a pandemic which are out of control of the organization.

3.0 Responsibilities

Overall - General Manager

Direct - Factory Manager

- Factory Officer
- Quality Officer/Food Safety Team Leader
- Administration Officer
- Assistant Factory Officer - Withering
- Assistant Factory Officer - Rolling
- Assistant Factory Officer - Firing
- Assistant Factory Officer - Sifting

4.0 Activities

4.1 Fire

4.1.1 Prevention and control of fire: Action shall be taken as per the Health and Safety Management System (Fire assessments, smoke, and fire alarm system, Fire hydrant systems, fire extinguishers, fire fighting teams, first aid teams) to prevent and control the incidence of fire in company premises.

4.1.2 Pre-prepared incident teams and fire drills and training shall be carryout annually.

4.1.3 In case of a fire or other type of emergency events following steps shall be followed with specifically trained and experienced people assembling as incident teams based on pre-prepared teams (Call Fire Brigade on the verge of fire: 091 2244445)

4.1.3.1 Activate the fire alarm system and switch off power from the main isolator.

4.1.3.2 Move everyone, including visitors, to the emergency assembly points.

4.1.3.3 Conduct a headcount and search for missing people.

4.1.3.4 Provide first aid for injured and transport to hospital if necessary.

4.1.3.5 Move all vehicles away from buildings without blocking the main entrances.

4.1.3.6 Take charge of office and telephones.

4.1.3.7 Stop outsiders from entering into the premises.

4.1.3.8 Visit the site and assess the damage to humans, property, and the environment.

4.1.3.9 Major accidents to be informed to top management and chairman.

4.1.3.10 Inform local police if necessary (XXXX Police Station, Tel No: 091 2222222).

4.1.3.11 Call the insurance and inform them of financial losses and nature damage.

4.1.3.12 Conduct an inquiry and prepare a report and assess the loss.

4.1.3.13 Organize repairs and refurbishments if necessary.

4.1.4 Possible hazards due to fire

4.1.4.1 Carbon monoxide, Carbon dioxide, Hydro chlorides, persistent organic pollutants, and carbon are chemical contaminants possible due to fire. Contamination of raw materials, in-process materials, packaging materials, and finished products with toxic gases produced during fire shall be checked before further processing.

4.1.4.2 Debris of buildings, plants, and other structures are possible physical contaminants due to fire. In the occurrence of such incidents, the raw materials, in-process materials, packaging materials, and finished products shall be checked before further processing.

4.1.4.3 Contamination of raw materials, in-process materials, packaging materials, and finished products with materials use for fire extinguishers can cause a chemical hazard. Biological hazards can be caused by the water used in fire extinguishing. All affected products shall be checked before further processing.

4.1.5 Any raw material, in-process product, or finished product affected by the fire shall be inspected and tested for suitability before use or dispatch them (FSMS 02/FSP/07 - Handling of potentially unsafe products).

4.1.6 Any production area of the plant or storage affected by fire shall be cleaned, repaired, and inspected before the commencement of operations.

4.2 Sabotage

4.2.1 All raw, packing, in-process, and finished products are kept under supervision, and access to materials shall be restricted to authorised personnel. Access to machinery and equipment also shall be restricted only to authorised personnel.

4.2.2 **Possible hazards due to sabotage** - Chemical contaminants such as diesel, kerosene oil, lubricant oil, detergent, and excessive quantities of materials such as chlorine, fertilizer can cause chemical hazards. All affected products shall be checked before further processing.

4.2.3 If any suspicious situation arises, all products and material suspected to be contaminated shall be segregated and checked for suitability of use with respect to contaminants. No products shall be issued until the safety of suspected products is confirmed.

4.3 Vehicle Accidents

4.3.1 If a container truck carrying made tea/raw material meets with an accident, the vehicle crew shall communicate the situation to the transport department/ administrative officer.

4.3.2 Depending on the situation, a replacement container truck with adequate capacity shall immediately be hired to the place of accident for recovery of made tea.

4.3.3 Incident plan

4.3.3.1 All injured parties shall be evacuated to the nearest hospital or for medical attention as necessary (If necessary, call for an Ambulance).

4.3.3.2 Douse, if any flames are observed from friction caused due to the collision of two vehicles.

4.3.3.3 Sealed off the accident incident area.

4.3.3.4 Damaged and thrown-out products or raw materials shall be collected into separate containers.

4.3.3.5 Inform the local police and call the fire brigade or towing vehicle for help if necessary.

4.3.3.6 Call insurance and inform the incident and damages.

4.3.3.7 On completion of police, investigations return all the products or raw materials to the factory for assessment of critically of damages.

4.3.3.8 Inform the management.

4.3.3.9 Assess the loss of material and vehicle.

4.3.3.10 Conduct an inquiry and prepare a report and assess the loss.

4.3.3.11 Organize repairs and refurbishments if necessary.

4.3.4 The breakdown/accident can cause an increase in product moisture if the packaging is damaged. A significant increase in moisture (above 25%) can cause biological hazards.

4.3.5 The recovered products shall be transported back to company storage and stored separately. The products shall be inspected and treated as per the procedure for handling of potentially unsafe products.

4.3.6 If a green leaf transporting truck meets with an accident, a recovery vehicle shall be hired or deployed immediately, and if the green leaves are affected by any contaminant or physical damages, such raw materials shall be segregated and discarded.

4.4 In the case of public emergency situations such as natural disasters, public health emergencies, or any such situations, adhere to the national disaster management centre guidelines and advice in the first place. Closedown factory operations and clean the facility before closing and after reopening as appropriate.

4.5 The environmental accidents, bioterrorism, workplace accidents other than those explained above shall be considered case by case basis where FST and top management should form as a task force in such situations while appointing relevant experienced employees to the assigned activities. Call for external help if necessary, use FSP 07, where necessary to handle affected products or material.

4.6 If the interruption of essential services such as water, electricity is taken place, switch into the internal power supply (generator) installed within the factory as given instructions, where company shall build additional water storage facilities to store clean portable water in a hygienic manner well enough for the operation to complete the ongoing batches of production and sanitizing the facility before interrupting the production in such case until service is back to the normalcy.

5.0 References

5.1 Procedure for the handling of potentially unsafe products – FSMS 02/FSP/07

5.2 Emergency control guide – FSMS 03/WI/EG/08

6.0 Records

6.1 Training and testing on emergency situations – FSMS 04/FSP/05/F/01

6.2 Emergency drills – FSMS 04/FSP/05/F/02

6.3 Reports and records on actual emergency situations – FSMS 04/FSP/05/F/03

ISO 22000 FOOD SAFETY PROCEDURE

FSP 06 – CORRECTION AND CORRECTIVE ACTION

Distribution:

1. Director/General Manager
2. Factory Manager
3. Factory Officer
4. Quality Officer
5. Assistant Factory Officer – Withering
6. Assistant Factory Officer – Rolling
7. Assistant Factory Officer – Firing
8. Assistant Factory Officer – Sifting
9. Consultant – FSMS

1.0 Objective

The objective of the procedure to control the nonconforming product been continued in the process and prevent the recurrence and occurrence of the defect while ensuring the critical limits for CCPs are exceeded, or when there is a loss of control for OPRPs, there are procedures in place to identify and control the use and release of the affected end products.

2.0 Scope

The scope covers all raw materials in stores, in-process materials, and finished products. The procedure also applicable to the corrective action for customer complaints, nonconformities related to products, processes, and food safety system.

3.0 Responsibilities

Factory Manager – Authorization and handling correspondence

Factory Officer – Coordination and evaluation

Quality Officer – Evaluation and execution

Assistant Factory Officers – Execution

4.0 Activities

4.1 Correction

4.2.1 When a loss of control of a CCP or an OPRP is discovered, or monitoring results indicate that the critical limit exceeded for CCP or OPRP, products shall be segregated and tagged for identification.

4.2.2 The responsible person shall take action to eliminate the nonconformity where each CCP and OPRP shall have predefined corrections associated with it for known failures.

4.2.3 The products segregated shall be handled according to the procedure for control of potentially unsafe products (FSMS 02/FSP/07).

4.2.4 The FST shall establish, maintain and update documented information, including method of identification, assessment, correction carried out for affected products to ensure their proper handling and arrangements for review of the corrections carried out.

4.2.5 If an OPRP action criterion is not met, identify the affected product lots and segregate them with tags, then determine the cause(s) of failure with consequences with respect to food safety. The reprocess products shall be destroyed/discarded if the nonconformity cannot be eliminated or discard. The results of the evaluation shall be retained as documented information.

4.2.6 The FST shall retain documented information describing corrections taken on nonconforming products and processes with the nature of the nonconformity and root cause(s) of the nonconformity and consequences as a result of the nonconformity with traceability information related to the lots of nonconforming products.

4.2.7 The correction made with revalidation, evaluation, and monitoring results shall also be recorded.

4.2 Corrective action

4.2.1 If monitoring results indicate that critical limits of CCP are exceeded or loss of control of OPRP, corrective action shall be initiated.

4.2.2 To eliminate the cause of the nonconformity or any other undesirable situation, corrective actions shall be taken.

4.2.3 The Factory Officer and the Quality Officer shall work with the Consultant – FSMS to identify the possible root causes and remedies to prevent recurring or occurring of the same incident again, and immediate action shall be taken to implement the solution, which will be evaluated and recorded.

4.2.4 Detail analysis shall be carried out to identify the root course through review of nonconformities identified by the customer and/or consumer complaints/ regulatory inspection reports related to the incident while reviewing trends in monitoring results that may indicate loss of control to determine the cause(s) of nonconformities.

4.2.5 The resulting action shall be implemented to ensure that nonconformities do not recur.

4.2.6 If the internal expertise is beyond the requirement, the above three persons responsible shall find a suitably knowledgeable expert to find remedies to the loss of control of the CCP or OPRP, who will be given limited authority to initiate corrective actions on a case by case basis.

4.2.7 The corrective action taken to eliminate the root course to prevent the recurrence of nonconformities shall be recorded. The corrective action taken shall be reviewed to evaluate the effectiveness, and results shall be recorded and retained as documented information.

4.3 Customer complaints

4.3.1 Customer complaints received verbally or in writing shall be forward to, General Manager using customer complaint form FSMS 04/FSP/06/F/01; wherever practical, together with the product, which shall be analysed and recorded in the Product Complaint File.

4.3.2 Factory Manager shall get the product through the auctioneer or through any other channel while making relief action for complaint immediately. The product shall be analysed to see whether it is defective. If it is defective, then the root cause for the nonconformity has to be determined.

4.3.3 Corrective action to be taken shall be communicated to person/s concern using corrective action report FSMS 04/FSP/06/F/02. The person responsible shall implement the corrective action. The effectiveness of the corrective action shall be analysed and recorded in the corrective action report once the corrective action is implemented (Use form for corrective action report).

5.0 References

5.1 Food Safety procedure for handling of potentially unsafe products

5.2 Hazard Control Plan - FSMS 02/HC/PL/01

5.3 Hazard Control Record - FSMS 04/HC/01/RC/01-04

5.4 Customer complaint form - FSMS 04/FSP/06/F/01

5.5 Corrective action report - FSMS 04/FSP/06/F/02

D. Ref: FSMS/04/FSP/06/F/01

XYZ TEA FACTORY

CUSTOMER COMPLAINTS / COMMENTS FORM

Please complete this form to help us to serve you better. All complaints/comments will be investigated to improve our products.

Name : .... Tel No: ....

Address : .... E-Mail: ....
....
....

Checklist to be marketed:

Product Details	**Made Tea**
1.Variety / Type	
2.Pack size	
3.Batch no:	
4.Date of expiry	
5.Storage condition	
6.Other comments	

Defects were observed by customer:...

Auctioneer's details:

1. Name & address: ....
2. Auctioneer's comments: ....

Buyer's details:

1. Name & address:..
2. Buyer's comments:………...

Product details:

Package		Taste		Colour		Appearance		Odour (Smell)		Moisture		Other
Opened		normal		normal		normal		normal		normal		
Closed		odd		differ		differ		differ		watery		

X – Mark as appropriate

To: General Manager

From: Buyer (Signature)...............................(Through) Auctioneer Date: ....

D. Ref: FSMS/04/FSP/06/F/02

XYZ TEA FACTORY

CORRECTIVE ACTION REPORT

Date: ....

To: .... From: ....

Type of nonconformity:

1. ……………………………………………………………..
2. ……………………………………………………………..
3. ……………………………………………………………..
4. ……………………………………………………………..

Cause for nonconformity:

....
....
....

Corrective action:

....
....
....

Confirmation of implementation:

....
....
....

Name:……………… Designation:………………… Signature:………………..

Effectiveness of corrective action:

....
....
....

Name:……………… Designation:………………… Signature:………………..

ISO 22000 FOOD SAFETY PROCEDURE

FSP 07 – HANDLING OF POTENTIALLY UNSAFE PRODUCTS

Distribution:

1. Director/General Manager
2. Factory Manager
3. Factory Officer
4. Quality Officer
5. Assistant Factory Officer – Withering
6. Assistant Factory Officer – Rolling
7. Assistant Factory Officer – Firing
8. Assistant Factory Officer – Sifting
9. Consultant – FSMS

1.0 Objective

The objective of the procedure is to prevent a nonconforming product being entering to food chain without ensuring that the food safety hazard of concern has been reduced to defined acceptable Levels.

2.0 Scope

The scope covers all the products of the food chain from raw materials after acceptance till the finished products reach the consumer.

3.0 Responsibilities

Overall – General Manager (Evaluation and authorization)

Direct – Factory Manager
- Factory Officer
- Quality Officer
- All AFOs (Identification segregation and communication to general manager)

4.0 Activities

4.1 If a product is found to be unsafe due to exceeding the critical limits or due to loss of control of OPRP or due to any other reason, the product has to be segregated and kept under hold condition by the head of the department.

4.2 Handling nonconforming products

4.2.1 When detection of nonconforming product is made, such person shall inform his Officer in-charge /Factory Manager.

4.2.2 The superior shall complete the relevant part of the form FSMS 04/FSP/07/RC/01 and notify them by sending it to the Factory Manager.

4.2.3 Suspected product (on visual inspection) shall be segregated and kept under "Yellow tag" if practical.

4.2.4 Factory Manager shall review the problem in cooperation with relevant section/department heads.

4.2.5 Factory Manager shall complete the relevant part of the form FSMS 04/FSP/07/RC/01 based on own review and send for action.

4.2.6 The recommendation shall be to rework to meet specified requirements or rejects as recommended in form FSMS 04/FSP/07/RC/01

4.3 Based on the above recommendations, the affected lot shall be checked for suitability. If suitable, products can be released for the next process. The products that are released to the food chain shall have evidence other than from the monitoring system that the control measures have been effective and the combined effect of control measures produces the intended performance.

4.4 Products affected by a failure of OPRPs shall only be released as safe when evidence other than the monitoring system demonstrates that the control measures have been effective and the combined effect of the control measures for that particular product conforms with the performance intended (i.e., identified acceptable levels) with the results of sampling, analysis and/or other verification activities demonstrate that the affected products conform with the identified acceptable levels for the food safety hazard(s) concerned.

4.5 If products have been released to the auction or market and are subsequently determined to be unsafe, the General Manager shall notify relevant interested parties and initiate a withdrawal/recall as to the FSMS 02/FSP/08.

4.6 The results of sampling, analysis, or other verification activities shall demonstrate that the affected product meets the identified acceptable level for the hazard. After implementation of the correction, the affected lot shall be re-evaluated for suitability to release.

4.7 If the hazard levels are not within limits, the possibility of achieving the acceptable levels through reprocessing or any other means shall be evaluated.

4.8 The controls and related responses from relevant interested parties and authorization for dealing with potentially unsafe products shall be retained as documented information.

4.9 If the hazard levels cannot be reduced up to a defined acceptable level, the product shall be disposed of as waste (PRP 10 - Work Environment).

4.10 Disposition of nonconforming product (rejects) shall be carried out under the supervision of the Department Manager to ensure the product is destroyed beyond reuse and prevent cross-contamination when disposed of through an incinerator or other method.

4.11 The completed nonconforming product and wastage report (if required) shall also be investigated for corrective and preventive action as per FSMS 04/FSP/07/RC/01.

4.12 Activities described in 4.2, 4.3, 4.4, 4.5, 4.6, and 4.7 have to be authorized by General Manager.

5.0 References

5.1	Hazard Control Plan	– FSMS 04/HC/PL/01
5.2	CCP Records	– FSMS 04/HC/01/RC/02
5.3	OPRP Records	– FSMS 04/HC/01/RC/01
		– FSMS 04/HC/01/RC/03
		– FSMS 04/HC/01/RC/04
5.4	PRP 10	– Work Environment

6.0 Records

6.1 Nonconforming Product and Wastage Report – FSMS 04/FSP/07/RC/01 are maintained by the Factory Manager and relevant section/ department head in the relevant file.

D. Ref: FSMS 04/FSP/07/RC/01

XYZ TEA FACTORY
Nonconforming Product and Wastage Report

Date:..

Name of product/Raw material: .... Quantity:......................................

To: FM/Through .... (Dept. Officer-in-charge) (Signature):..................................

From: .... (Name/Designation):.................... (Signature): ……………………..

Nonconformity :

....

....

Probable cause:

....

....

Lot No. /Batch No.: ....

Location: ....

To: ....

Through:..

From: Factory Manager: .... (Signature): ……………… Date:..................

I recommend following actions (rework/re-inspect/reject/scrap/disposed)

....

....

Please implement the above and return this form in conformation (Keep one copy in relevant dept.)

Required to submit/not submit to management review meeting

To: FM/Through..................... (Dept. Officer-in-charge) (Signature): ....

From: .... (Name/Designation):.................. (Signature): ……………………..

I conform the implementation of above action (rework/re-inspect/reject/scrap/disposed)

Date implemented:………………………………………

ISO 22000 FOOD SAFETY PROCEDURE

FSP 08 - WITHDRAWALS / RECALLS

Distribution:

1. Director/General Manager
2. Factory Manager
3. Factory Officer
4. Quality Officer
5. Assistant Factory Officer - Sifting
6. Consultant - FSMS
7. Auctioneer

1.0 Objective

The objective is to prevent reaching consumers, a product identified as unsafe after dispatch.

2.0 Scope

The scope covers all the varieties of made tea stated under the scope of the food safety management system.

3.0 Responsibilities

General Manager

Authorizing the withdrawal and allocation of resources

Factory Manager

Analysing the requirement for withdrawal and recommending the needs for withdrawal communicating via E-mail to manager stores at auctioneer's stores with a copy to the head of the auction agency for withdrawal from their storage, to Sales Administration Manager or Marketing Manager requesting to copy to buyer, thereby requesting for withdrawal from distributors and or from retail outlets.

Quality Officer

Holding the product/batch due to complaints received from the market. Analyse the situation in market and auction, communicating the finding to Factory Manager to analyse the need for withdrawal.

Auctioneer

Communicate to warehouse manager and buyer for withdrawal process from buyer, distributors, or retailers in the specific market.

The resources for withdrawal shall be arranged communicating with General Manager.

Factory Officer/Assistant Factory Officer – Sifting

Identification of records of relevant batch/product while furnishing the number of units and weight of the batch.

Execute the withdrawal from storage.

4.0 Activities

4.1 Requirement for withdrawals/recall

When a complaint has been received, the product and/or stocks in particular stores and/or entire batch shall be held based on the nature of the complaint. If a particular product is found contaminated and not suitable for consumption, then it is necessary to evaluate whether it is limited to a single pack or is common to the entire batch by analysing more samples. If it is common to the entire batch, all units of the particular batch shall be withdrawn/recalled.

4.2 Identification of the batch needs to recall/withdrawal.

The batch no, date of manufacture, pack size, and the verity must be accurately identified. The Factory Officer shall furnish the total number of units produced for this particular batch. Use FSMS 04/FSP/07/RC/01.

4.3 Communication to relevant personal

Holding the product/s or batch/s shall be carried out based on the findings by Factory Officer. Factory Manager shall inform Auctioneer and Stores Manager at auctioneer's stores, the batch number, date of manufacture, and the no of units to be withdrawn. The withdrawal/recall shall be authorized by General Manager. The communication regarding withdrawal/recall from buyer or distributor stores or from retail outlets shall be done through the Auction Manager. On-site verification regarding defective product/s, batch/s shall be done by Factory Officer.

The information shall be communicated to the general public via mass media as and when necessary.

4.4 Execution of withdrawal/Recall

4.4.1 Auctioneer's Storage

Quality Officer shall execute the withdrawal/recall from storage and prevent distribution from storage outside the premises.

The duration for completion of the withdrawal/recall and the number of units withdrawn/ recalled shall be monitored and recorded.

4.4.2 Exported Products

Communication regarding withdrawal/recall for export products shall be done through the Auction Manager. He shall communicate details of products/batches to be withdrawn/recall to the Buyers/agents who export the items. The items shall be disposed of at respective locations, and confirmation of the number of units shall be collected.

4.5 Disposal of products

The withdrawn/recalled products shall be stored separately with identification to prevent any unintended use and cross-contamination. The withdrawn/recalled products shall be disposed of/reprocessed as per the instruction given by the

Factory Manager (FSMS 04/FSP/07/RC/01). The implementation of the recommended action shall be confirmed by Factory Officer and Quality Officer.

4.6 Evaluation of the effectiveness of the withdrawal/recall

The effectiveness of the withdrawal/recall or practice withdrawal/recall shall be evaluated through a mock withdrawal at each step and retain documented information. The effectiveness of the withdrawal/recall after different time intervals of production shall be evaluated. The time duration and number of units received shall be monitored to evaluate the effectiveness.

4.7 The traceability of the finished product shall be used to evaluate the effectiveness of the withdrawals/recalls from distributor stores and from a retail outlet. All evaluations shall be monitored and recorded. The evaluation shall be carried out annually.

4.8 The actual or mock withdrawals shall establish and maintain documented information for notifying relevant interested parties (e.g., statutory/regulatory authorities, customers, and/or consumers), handling withdrawn/recalled products as well as products still in stock, and performing the sequence of actions to be taken.

5.0 References

5.1 FSP for Handling of Potentially Unsafe Product – FSMS 02/FSP/07

5.2 FSP for Correction and Corrective action – FSMS 02/FSP/06

5.2 FSP for Product Identification and Traceability – FSMS 02/FSP/04

6.0 Records

6.1 Annual mock withdrawals records – FSMS 04/FSP/08/F/01

6.2 Actual recalls and withdrawals – FSMS 04/FSP/08/F/02

6.3 Analysis reports of efficiency of mock or actual recalls and improvement strategies – FSMS 04/FSP/08/F/03

ISO 22000 FOOD SAFETY PROCEDURE

FSP 09 – INTERNAL FOOD SAFETY AUDITS

Distribution:

1. Director/General Manager
2. Factory Manager
3. Factory Officer
4. Quality Officer
5. Assistant Factory Officer – Withering
6. Assistant Factory Officer – Rolling
7. Assistant Factory Officer – Firing
8. Assistant Factory Officer – Sifting

1.0 Objective

To determine whether or not the food safety management system conforms to planned arrangements, requirements, and to ISO 22000:2018 standard and is being effectively implemented and updated continuously.

2.0 Scope

The scope covers all activities related to M/s XYZ Ltd.'s food safety management system.

3.0 Responsibilities

Overall – General Manager
Direct – Food Safety Team Leader
– Quality Officer
– Consultant – FSMS

4.0 Activities

4.1 Internal food safety audit plan

4.1.1 The audit plan is prepared annually by the Food Safety Team Leader/Consultant - FSMS, at the beginning of each year, showing where and when the audits will take place, based on the status and importance of the activity to be audited, and the scope. The form for the internal food safety audit annual plan is given in FSMS 04/FSP/09/F/01.

4.1.2 The internal audits shall be conducted every three months (quarterly) covering each and every department, with priority to if any food safety incidents happened during the last three months period while evaluating the performance of corrective actions taken and revalidation with evaluations conducted.

4.1.3 The audit plan shall also consider the importance of the processes concerned, changes in the FSMS, and the results of monitoring, measurements, and previous audits.

4.1.4 The food safety team leader shall define the audit scope for each individual department separately, with specific auditors assigned for the given audit highlighting specific concerns for the audit.

4.1.5 The audit team shall conduct the complete audit for the entire system annually to validate the system while conducting partial audits with relevant importance as the internal audit program.

4.1.6 The external compliance audits shall be conducted by relevant independent certification providers by every six months as to the signed contracts.

4.2 Audit schedule and notice

Notice of the internal audit shall be sent to the section head one (01) week before the audit date. The scope & schedule shall be given in the notice (FSMS 04/FSP/09/F/02).

4.3 Internal food safety audit team

4.3.1 The internal food safety audit team shall comprise at least two members of the audit team. The team leader shall be appointed based on knowledge and experience.

4.3.2 To ensure the objectivity and impartiality of the audit process, auditors shall not be nominated to audit their division or department.

4.4 Auditing

The team leader shall plan and carry out the audit with other audit members according to the scope given in the notice.

4.5 Guidelines for the conduct of audits, reports, and follow up

4.5.1 Audits should be conducted using the following guidelines:

a. Do not include the Manager or Officer responsible for the department to be audited.
b. Give sufficient time between notice and the audit, whenever possible.
c. Examine past audits, non-conformances, and corrective action requests.
d. Controlled documents such as Standards, PRPs, FSPs, or WI/SOPs shall be used as references or standards for auditing.
e. Focus on constructive issues.
f. Present the report to the persons responsible for any corrective actions.

4.5.2 The non-compliances shall be categorized into Critical, Major, or Minor nonconformities based on the following definitions:

4.5.2.1 *Critical Deficiencies*

Critical deficiencies are which:

a. Indicate a significant risk that a product could or would be harmful to the consumer.
b. Have previously been documented as a critical or major deficiency by a licensing authority.
c. Would result in adulteration or miss-branding of finished products
d. Violation of critical control point or operational prerequisite program
e. Would resulting from the product not meeting final product specifications.

4.5.2.2 *Major Deficiencies*

A Major deficiency is a non-critical deficiency which:

- Could or would produce a product, which is not in accordance with the products, registration requirement.
- Indicates a significant deviation from a clause of the prerequisite program, which may result in:
  a. Finished product not meeting written specifications
  b. Product mix-up
  c. Product miss-formulation
  d. Cross-contamination, microbiological contamination, or adulteration
  e. Failure of the product to meet shelf-life specifications
  f. Contravenes of written instructions or procedures without proper justification or authority.
  g. Allows loss of process control so that the process outcome is not consistent or predictable.

4.5.2.3 *Minor Deficiencies*

Other deficiencies cannot be classified as either critical or major, but which:

a. Indicate a departure from any clause of the ISO 22000 standards
b. Contravene written or accepted M/s XYZ Ltd., tea factory standards.
c. In the opinion of the auditor may lead to Major or Critical deficiencies under certain circumstances.

   NOTE: *The sum of several other deficiencies that together indicate a lack of control over the system should be combined and incorporated as a Major deficiency.*

d. The audit team should prepare, if necessary, a full audit report containing observations, elaborations, and audit team recommended corrective actions. This report should not be circulated with the audit report form.
e. Follow-up shall take the form of presentation of an audit report, agreement on corrective action with a schedule, and verification by the auditor that the corrective action has been completed. Audit reports shall be numbered and stored by the FSTL.

4.6 Reporting results

All audit findings shall be discussed and raised non a conformity report (form FSMS 04/FSP/09/RP/01), if any. The team leader shall write a summary report (form FSMS 04/FSP/09/RP/02) giving reference to the samples selected, the areas find with a lot of weaknesses, etc.

4.7 Corrective action

The person responsible for the area where nonconformity is found shall determine and record the corrective action defining the duration required. The corrective action shall be completed within the stipulated time.

4.8 Follow up audits

When the corrective action is completed, a follow-up audit shall be carried out. Completion and effectiveness shall be verified during follow-up audits. The results shall be recorded in the relevant section of NCR (FSMS 04/FSP/09/RP/01)

4.9 Compliance Audits

In the conduct of compliance audits, the responsible auditor shall include the compliance status of the Food Safety Policy, Objectives, relevant Food Safety Procedures, Prerequisite Programs Work Instructions, and Standard Operating Procedures with testing and monitoring documents or areas as part of the audit(s).

4.10 Documented information

The results of the audits, corrective action, follow-up audits, and results of the compliance audits shall be retained as documented information as evidence of the implementation of the audit plan and the audit results

5.0 References

5.1	Internal Food Safety Audit Plan	- FSMS 04/FSP/09/F/01
5.2	Internal Food Safety Audit Notice	- FSMS 04/FSP/09/F/02
5.3	Internal Food Safety Audit Format	- FSMS 04/FSP/09/F/03
5.3	Internal Nonconformity Report	- FSMS 04/FSP/09/RP/01
5.4	Internal Food Safety Audit Summary Report	- FSMS 04/FSP/09/RP/02
5.5	FSMS Validation Guide	- FSMS 03/WI/VG/07

6.0 Records

6.1	Internal Food Safety Audit Records	
6.2	FSMS System Verification Check List	- FSMS 04/FSP/09/CL/01
6.3	FSMS Validation Check List	- FSMS 04/FSP/09/CL/02
6.4	Internal Nonconformity Report	- FSMS 04/FSP/09/RP/01
6.5	Internal Food Safety Audit Report	- FSMS 04/FSP/09/RP/02
6.6	Pest Control Check List	- FSMS 04/PRP/07/CL/03

D. Ref: FSMS 04/FSP/09/F/01

XYZ TEA FACTORY

INTERNAL FOOD SAFETY AUDIT PLAN

For the year ....

To:.. Section: ....

From: .... Date. : ....

Quarter	Month	Dept./Division	Scope

D. Ref: FSMS 04/FSP/09/F/02

XYZ TEA FACTORY

INTERNAL FOOD SAFETY AUDIT NOTICE

To:...

From:..

Section: ....

Date. : ....

Audit scope:

1. ....
2. ....
3. ....
4. ....

Audit Schedule:

Date :...

Time : ....

Audit Team : 1) ....

2) ....

3) ....

....
Signature
Position of the FSTL

D. Ref: FSMS 04/FSP/09/RP/01

XYZ TEA FACTORY
INTERNAL FOOD SAFETY AUDIT CORRECTIVE ACTION REQUEST (CAR) FORM

Non Conformity Record			
Date:		NCR No:	
Department:		Sheet No:	
Auditor:		Designation:	
Auditee:		Designation:	
Process Audited:			
Audit Scope:			

Non Compliance Observed:			
Description:			
Reference: Standard Clause/Intensity			
Root Cause Analysis:			
Proposed Corrective Action:			
Proposed by:		Date of Compliance:	
Auditor Signature:		Auditee Signature:	

Follow up Audit Date:		Date of Completion:	
Verification of Corrective Action:			
Auditor Signature:		Auditee Signature:	
Comments If Any:			

D. Ref: FSMS 04/FSP/09/RP/02

XYZ TEA FACTORY
INTERNAL FOOD SAFETY AUDIT SUMMARY REPORT

Dept. /Division: ....
Date: ....

Audit Scope:
....
....
....
....
....

Audit team leader's Comments:
....
....
....
....
....
....
....
....
....
....
....
....
....
....
....
....
....
....
....

Auditor Name (Leader):.................................. Signature : ....

Auditor Name:.. Signature : ....

Auditor Name: .... Signature : ....

User Guide

The ISO 22000 food safety management system requirements are validated according to the standard, with a YES or No direct answers and a description with YES answers to evaluate if there is any implication to the food safety and whether any changes needed to the ISO 22000 FSMS to adjust the current state of practice to eliminate a given food safety implication. The food safety team is free to decide the answers within given parameters, which will be easy to understand and find the nonconformities quickly.

Validation Type (check one):

☒ Initial Validation (within 12 months of implementation)

☐ Validation (Reassessment) due to changes made in raw materials or source of raw materials; product formulation; processing methods or systems, including computers and their software; packaging; finished product distribution systems; or the intended use or intended consumers of the finished product and rate or type of consumer complaints.

☐ Annual Validation (Reassessment) of the ISO 22000 (with Hazard Control Plan, including CCPs, OPRPs, Hazard Analysis, and all PRPs)

Date Conducted :

Conducted By :

Designation :

System Validation					
Topic	YES	NO	If "YES" Describe	Food Safety Implication	Are Modifications to the FSMS required?
1. General					
Any changes to interested parties or customers?	☐	☐			
Any specific changes to regulatory, statutory customer requirements?	☐	☐			
Any changes to the processing facility or infrastructure?	☐	☐			
Renovations or extensions to any existing facility?	☐	☐			
Any repetitive machine failures, breakdowns, or reported issues?	☐	☐			
Any pest infestations, flooding, or disastrous event in the recent past?	☐	☐			
Any frequent accidents, power failures, or essential service interruptions?	☐	☐			
	☐	☐			
2. Steps to Follow Before Validation					
Did the validation guide explain the updated validation methods based on the most current technological advances in the industry?	☐	☐			
All the methods of validation are in compliance?	☐	☐			
Did the critical control points, critical control limits, or observable action criteria used is up-to-date and well within requirements?	☐	☐			
Does FST have reviewed its validation methods and tested for compliances?	☐	☐			
Are the documented validation methods available for reference with results for verification?	☐	☐			
Check all the records related to each CCP, OPRP, and PRPs for any abnormalities	☐	☐			
Gather information on mock withdrawals or recalls fire drills or real events	☐	☐			
Training and workshops or updated knowledge on issues internal or external	☐	☐			
Any other	☐	☐			

System Validation

Topic	YES	NO	If "YES" Describe	Food Safety Implication	Are Modifications to the FSMS required?
1. Policy and Procedures					
Any changes to scope?	☐	☐			
Any changes to food safety policy?	☐	☐			
Any changes to food safety objectives?	☐	☐			
Any deviations occurred while achieving food safety objectives?	☐	☐			
Any changes to food safety procedures?	☐	☐			
Any changes to prerequisite programs?	☐	☐			
Any deviations observed for food safety procedures or PRPs?	☐	☐			
Any changes to a process flow diagram (s)?	☐	☐			
Any changes to site schematic diagrams?	☐	☐			
2. Evaluate product & process					
Product description changed, e.g., intended use, the consumer?	☐	☐			
Formula changed?	☐	☐			
Ingredients / Packaging changed?	☐	☐			
Any new product consumption or storage methods?	☐	☐			
Any new suppliers?	☐	☐			
Process flow changed?	☐	☐			
Equipment/computer software changed?	☐	☐			
Finished Product Distribution changed?	☐	☐			
Other, e.g., production volume increased :	☐	☐			

System Validation

Topic	YES	NO	If "YES" Describe	Food Safety Implication	Are Modifications to the FSMS required?
3. Evaluate product/process history					
Repeat CCP deviations?	☐	☐			
Any recent industry recalls of similar product since the last annual validation?	☐	☐			
New or emerging hazards, e.g., recent Salmonella Morbidity & Mortality problems identified with sausages.	☐	☐			
Regulatory agency recommendations, e.g., guidance documents, regulations?	☐	☐			
Any confirmed food safety consumer complaints?	☐	☐			
4. Evaluate the adequacy of CCPs, critical limits, monitoring, corrective action, CCP verification, OPRPs, and record-keeping procedures. Review current CCP documentation.					
Do the CCPs control the hazards?	☐	☐			
Are the CCP critical limits adequate?	☐	☐			
Do monitoring methods and frequency demonstrate control?	☐	☐			
Do corrective actions properly address affected products and correct deviations?	☐	☐			
Does validation include a review of consumer complaints?	☐	☐			
Are the OPRP control limits adequate?	☐	☐			
Do monitoring methods and frequency demonstrate control of OPRP?	☐	☐			
Any changes	☐	☐			
Do emergency preparedness and response monitoring documents/records adequate?	☐	☐			
Other, e.g., PRP or procedures may affect the hazard analysis :	☐	☐			

FSP Validation

Topic	YES	NO	If "YES" Describe	Food Safety Implication	Are Modifications to the FSMS required?
5. Evaluate the adequacy of FSPs					
Food Safety Procedure #1 - Control of Documents	☐	☐			
Are the FSP control measures up to date and adequate?	☐	☐			
Do monitoring methods and frequency demonstrate control?	☐	☐			
Food Safety Procedure #2 – Statutory, Regulatory & Customer Requirements	☐	☐			
Are the FSP control measures up to date and adequate?	☐	☐			
Do monitoring methods and frequency demonstrate control?	☐	☐			
Food Safety Procedure #3 - Communication	☐	☐			
Are the FSP control measures up to date and adequate?	☐	☐			
Do monitoring methods and frequency demonstrate control?	☐	☐			
Food Safety Procedure #4 - Product Identification & Traceability	☐	☐			
Are the FSP control measures up to date and adequate?	☐	☐			
Do monitoring methods and frequency demonstrate control?	☐	☐			
Food Safety Procedure #5 - Emergency Preparedness & Response	☐	☐			
Are the FSP control measures up to date and adequate?	☐	☐			
Do monitoring methods and frequency demonstrate control?	☐	☐			
Food Safety Procedure #6 - Corrective Action & Correction	☐	☐			
Are the FSP control measures up to date and adequate?	☐	☐			
Do monitoring methods and frequency demonstrate control?	☐	☐			
Food Safety Procedure #7 – Handling of Potentially Unsafe Products	☐	☐			
Are the FSP control measures up to date and adequate?	☐	☐			
Do monitoring methods and frequency demonstrate control?	☐	☐			

FSP Validation

Topic	YES	NO	If "YES" Describe	Food Safety Implication	Are Modifications to the FSMS required?
Food Safety Procedure #8 – Withdrawal/Recall	☐	☐			
Are the FSP control measures up to date and adequate?	☐	☐			
Do monitoring methods and frequency demonstrate control?	☐	☐			
Food Safety Procedure #9 – Internal Food Safety Audits	☐	☐			
Are the FSP control measures up to date and adequate?	☐	☐			
Do monitoring methods and frequency demonstrate control?	☐	☐			
Food Safety Procedure #10 – Management Review	☐	☐			
Are the FSP control measures up to date and adequate?	☐	☐			
Do monitoring methods and frequency demonstrate control?	☐	☐			
Food Safety Procedure #11 -	☐	☐			
Are the FSP control measures up to date and adequate?	☐	☐			
Do monitoring methods and frequency demonstrate control?	☐	☐			
Any additional info related to FSP					
	☐	☐			
	☐	☐			
	☐	☐			
	☐	☐			
	☐	☐			
	☐	☐			
	☐	☐			

PRP Validation

Topic	YES	NO	If "YES" Describe	Food Safety Implication	Are Modifications to the FSMS required?
6. Evaluate the adequacy of PRPs					
Prerequisite Program #1 – Construction, layout, and equipment	☐	☐			
Are the PRP control limits adequate?	☐	☐			
Do monitoring methods and frequency demonstrate control?	☐	☐			
Prerequisite Program #2 – Management of purchased materials	☐	☐			
Are the PRP control limits adequate?	☐	☐			
Do monitoring methods and frequency demonstrate control?	☐	☐			
Prerequisite Program #3 – Preventive maintenance	☐	☐			
Are the PRP control limits adequate?	☐	☐			
Do monitoring methods and frequency demonstrate control?	☐	☐			
Prerequisite Program #4 – Control of monitoring and measuring equipment and methods	☐	☐			
Are the PRP control limits adequate?	☐	☐			
Do monitoring methods and frequency demonstrate control?	☐	☐			
Prerequisite Program #5 – Storing, handling, and transportation of material	☐	☐			
Are the PRP control limits adequate?	☐	☐			
Do monitoring methods and frequency demonstrate control?	☐	☐			
Prerequisite Program #6 – Personnel hygiene	☐	☐			
Are the PRP control limits adequate?	☐	☐			
Do monitoring methods and frequency demonstrate control?	☐	☐			
Prerequisite Program #7 – Pest control	☐	☐			
Are the PRP control limits adequate?	☐	☐			
Do monitoring methods and frequency demonstrate control?	☐	☐			

PRP Validation					
Topic	YES	NO	If "YES" Describe	Food Safety Implication	Are Modifications to the FSMS required?
Prerequisite Program #8 – Sanitary and employee facilities	☐	☐			
Are the PRP control limits adequate?	☐	☐			
Do monitoring methods and frequency demonstrate control?	☐	☐			
Prerequisite Program #9 – Supply of air, water, and energy	☐	☐			
Are the PRP control limits adequate?	☐	☐			
Do monitoring methods and frequency demonstrate control?	☐	☐			
Prerequisite Program #10 – Work environment	☐	☐			
Are the PRP control limits adequate?	☐	☐			
Do monitoring methods and frequency demonstrate control?	☐	☐			
Prerequisite Program #11 – Cleaning and sanitisation of plant, machinery, equipment, and utensils	☐	☐			
Are the PRP control limits adequate?	☐	☐			
Do monitoring methods and frequency demonstrate control?	☐	☐			
Prerequisite Program #12 – Supplier Evaluation & Selection	☐	☐			
Are the PRP control limits adequate?	☐	☐			
Do monitoring methods and frequency demonstrate control?	☐	☐			
Prerequisite Program #13 – Food Defence	☐	☐			
Are the PRP control limits adequate?	☐	☐			
Do monitoring methods and frequency demonstrate control?	☐	☐			
7. Any other:					
Add rows as necessary	☐	☐			

Each FSP and PRP shall be tested for its application, the correctness of procedure, and its validity in the current context of the organization. Each FSP and PRP also need to be validated with its recording structure and if all the retrieved information regards to the controls or observations are accurate and up-to-date with the relevant performance criteria set for evaluation. If there is any violation, mark it.

User Guide

The ISO 22000 food safety management system requirements are assessed according to the standard, with a YES or No direct answers and a numerical value to evaluate the actual state of the current practice. The internal auditor is free to decide the answers within given parameters, which will be easy to understand and find the nonconformities quickly.

Verification activities shall begin once the validation process is completed. The objective of verifications is to continuous monitoring of process that critical food safety attributes shall be controlled throughout the process, where FST shall define the key indicators.

These results shall be statistically analyzed to identify important trends and actions that need to be taken to build a more robust FSMS through continuous improvement. Verification can answer key questions, which are summarized below.

Each requirement identified below in the table will be assessed on the following basis.

0 – No compliance with the criteria
1 – Comply slightly with the criteria
2 – Comply partly with the criteria
3 – Comply fully with the criteria

Identify areas that are not complying with the give PRP parameters and mark them accordingly in the special remarks column as follows:
Mark **MN** for **Minor concerns**: Those which can be corrected without the need for physical/financial investments from the company.
Mark **MA** for **Major concerns**: Those which are critical to the success of the system and require physical/financial investments.

In addition, if the auditor feels to write something on the matter, he still can use the same column or an additional paper. However, he has to write the audit report with the collaboration of the audit team leader.

Internal Audit – System Verification

ISO 22000 Food Safety Requirements	Yes	No	Score	Comments
Context of Organization				
Have you identified, reviewed, updated documented context of the organization, including internal and external issues related to food safety of orthodox black tea manufacturing?				
Have you defined interested parties, including applicable statutory, regulatory, and customer requirements with regard to food safety?				
Have your FST established, implemented, and maintain a procedure to identify and update to understand the needs and the expectations of interested parties.				
Have the products, processes, and production sites that will be addressed by the food safety management system been identified?				
Have your FSMS defined and documented your place in the food supply chain and the scope of the operation?				
Leadership				
Does your organization have clearly defined its food safety policy and objectives?				
Does your organization have set up the FSMS to achieve these policies and objectives?				
Have you planned and implemented the food safety management system, and are you operating, maintaining, and updating that system?				
Does the organization have synchronized FSMS requirements into its business processes while allocating resources required and directing and supporting employees to contribute for effective implementation?				
Have you influenced relevant management roles to demonstrate their responsibility while promoting continual improvement as well as address internal/external communication?				
Does your food safety policy is available for all interested parties, and is it displayed inside the facility?				
Have you defined structure, roles, and responsibilities, as well as delegated authorities to competent persons?				
Have you appointed the food safety team and food safety team leader with necessary responsibilities and authority?				

Internal Audit – System Verification				
ISO 22000 Food Safety Requirements	Yes	No	Score	Comments
Planning				
Has FST identified risk and the opportunities to be addressed in order to achieve intended results while enhancing desirable effects and continual improvements as well as to prevent or reduce undesirable effects?				
Does the organization have planned actions to evaluate the effectiveness of the FSMS processes?				
What documented actions organization has taken to address risks and opportunities, and are they impactful?				
Do you have a consistent, measurable, monitored food safety policy that has been verified and communicated?				
Have you planned to achieve food safety objectives with relevant specific criteria in mind?				
How have you planned your process of changes, including purpose, continued integrity, and availability of resources?				
Support				
Have you considered using internal as well as external resources to develop and implement FSMS?				
Are you sure that the external experts you engage are competent and that records were kept defining the responsibility and authority of these experts?				
Are you using externally developed elements in your FSMS, and is it customised according to your requirements?				
Have you ensured that employees with food safety related jobs are aware of their responsibilities and competent?				
Do employees with multiple responsibilities clearly understand them and competent to carry them out effectively?				
Have employees with specialized needs obtained the required training?				
Is the infrastructure necessary for the establishment and maintenance of your FSMS in place?				
Have you provided the resources necessary for the establishment, management, and maintenance of the work environment required by the food safety management system?				
Has your organization established a mechanism for communicating with external parties about food safety issues?				
Has a member of the organization (e.g., the food safety team leader) been assigned responsibility for liaison with external organizations on matters related to food safety?				
Are records kept of external communications on food safety issues, and are the appropriate food safety requirements available or accessible?				
Does your organization have mechanisms to ensure that information on food safety, particularly about input or product changes and the introduction of new products, processes, or equipment, is provided to FST timely manner?				

Internal Audit – System Verification

ISO 22000 Food Safety Requirements	Yes	No	Score	Comments
Documented Information				
Has your organization documented its food safety management system?				
Does your organization have all the documented information required to demonstrate the effective development, implementation, and updating of its FSMS?				
Does your organization has implemented controls for documented information, including identification, description format, review and approval process, protections, authorisation, storage distribution access, and retrieval, etc.?				
Does your organization have a documented procedure concerning control of documented information?				
Operation				
Have you planned, implement, control, maintain, and updated the processes needed to meet requirements for the realization of safe products and with processes and controls in accordance with identified requirements?				
PRP				
Has a set of prerequisite programmes been established, implemented, and maintained?				
Has your organization recorded the process that is used to select its PRPs and documented how the activities included in the PRPs are managed?				
Have you used the applicable Technical Specification in the ISO/TS 22002 series or applicable standards, codes of practice, and guidelines to develop your PRPs?				
Treatability				
Does your organization has implemented a documented traceability system with evidence for verifiable results covering incoming material from the suppliers and the first stage of the distribution route of the end product?				
Can you trace back a given quantity of end products with the number of ingredients and suppliers as evidence of effectiveness?				
EPR				
Have you established a system to manage potential emergency situations and accidents that impact on food safety?				
Do you have documented information with relevant competent persons allocated with appropriate training on EPR?				
Does EPR has planned action to reduce the consequences of the emergency situation, appropriate to the magnitude of the emergency or incident and the potential food safety impact?				

Internal Audit – System Verification				
ISO 22000 Food Safety Requirements	Yes	No	Score	Comments
Hazard Control				
Has a description of all raw materials, ingredients, and product contact materials been documented?				
Have you documented the characteristics of your end products?				
Have all reasonably expected intended uses of the end product been identified and documented?				
Has your food safety team developed flow diagrams for each product or process category?				
Has the food safety team done an on-site check and made any required modifications to the flow diagrams?				
Has your food safety team prepared plant or site-specific schematics and identified the opportunities for cross-contamination?				
Has the team described and recorded the food safety control measures or procedures at each step in the process?				
Does your organization understand the concepts of prerequisite programmes, operational prerequisite programmes, and critical control points?				
Does your food safety team understand this process, and has it determined and documented the approach that it will use to make decisions?				
Do you understand what is involved in conducting a hazard analysis?				
Has a food safety team with the required knowledge and experience been set up?				
Has the food safety team identified all the reasonably expected food safety hazards for each step in your operation?				
Has the team determined, justified, and recorded the acceptable level for each hazard at each step in your operation?				
Has the team decided how to carry out its hazard assessment, determined whether or not elimination, control, or reduction of each hazard is required, and recorded the results of this assessment?				
Has your team selected a specific methodology (e.g., decision tree/risk matrix) that it will use for assessing and sorting the control measures it has identified into PRPs, operational PRPs, and CCPs, and has it documented its use?				
Has your team revised its documented PRPs subsequent to the completion of the hazard analysis?				
Has the food safety team documented each of the operational PRPs in your food safety management system?				
Has the food safety team documented each CCP in the hazard control plan?				
Has the team set a measurable critical limit for each CCP and provided the necessary instructions and support for its implementation?				

Internal Audit – System Verification				
ISO 22000 Food Safety Requirements	Yes	No	Score	Comments
Has a monitoring system been established for each CCP?				
Has the food safety team developed corrections and/or corrective actions for each CCP and provided training on these to the personnel involved?				
Has the food safety team updated all the information gathered in the preliminary steps?				
Control of Monitoring and Measuring				
Does your organization has documented how it controls monitoring and measuring methods, equipment in use, as well as the adequate monitoring of measuring activities related to hazard control plan and PRP?				
Do methods use to assess the validity of the previous measurement results when the equipment or process environment is found not to conform to requirements, and what actions will be taken in such situations?				
Does FST conduct analysis of the results of verification as a performance evaluation criteria?				
Validation and verification of the food safety management system				
Does your food safety team understand the differences between verification and validation and the concepts of improvement and updating?				
Has your food safety team validated the control measures (e.g., operational PRPs and CCPs) included in your organization's food safety management system?				
Has the food safety team planned verification activities for each control measure or set of control measures?				
Are the results of individual verification activities evaluated and the required follow-up actions taken?				
Does the food safety team regularly analyse the results of verification activities?				
Can your organization demonstrate that the measuring equipment and methods have been calibrated or verified and that the necessary adjustments and safeguards are in place?				
Control of Nonconformity				
Does your organization understand the importance of controlling nonconforming products?				
Has the food safety team identified correction(s) for each CCP and operational PRP?				

Internal Audit – System Verification

ISO 22000 Food Safety Requirements	Yes	No	Score	Comments
Has the food safety team established and documented procedures for the control of the causes of nonconformity?				
Does your organization assign responsibilities and authorities for evaluating monitoring procedures and initiating corrective actions?				
Does your organization have in place a procedure to secure the control of potentially unsafe product so that it does not enter the food chain prior to evaluation?				
Does your organization have in place a process for effectively evaluating products affected by nonconformity prior to their release into the food chain?				
Does your food safety management system provide for safe handling of the product that is not acceptable?				
Has top management appointed personnel with the authority and responsibility to initiate and execute withdrawals?				
Does your organization have procedures that enable and facilitate the withdrawal of lots of end products identified as either potentially or actually unsafe?				
Does your food safety management system include procedures for the control and proper handling of withdrawn products and for recording their outcomes?				
Are withdrawals reported to top management at the scheduled management review?				
Has your organization recently tested its withdrawal procedures?				
Performance Evaluation				
Do you have planned activities specifying what activities require monitoring, measurement analysis, and evaluation?				
Does your performance evaluation identify the need for updating and improving the FSMS or trends which indicate a higher incidence of potentially unsafe products or process failures?				
Do results of analysis and resulting activities be retained as documented information and reported to top management as input to the management review and the updating of the FSMS?				
Internal Audit				
Does your organization conduct internal audits at planned intervals according to documented procedures?				
Does your organization act in a timely way on nonconformities identified by the internal audit?				
Are your internal auditors trained, and are procedures in place to ensure that they are impartial and objective?				

Internal Audit – System Verification				
ISO 22000 Food Safety Requirements	Yes	No	Score	Comments
Management review				
Has your organization reviewed its food safety management system as planned?				
Does the information provided to top management permit them to effectively review the food safety management system?				
Do the records of the review demonstrate the organization's ongoing commitment to food safety and to the continual improvement of its system?				
Improvement of the food safety management system				
Can your organization's top management demonstrate that it is committed to the continual improvement of the food safety management system?				
Has the food safety team planned for and undertaken evaluations of the food safety management system, and has it updated the system to implement any needed changes?				
Any Additional Verifications				

ISO 22000 FOOD SAFETY PROCEDURE

FSP 10 - MANAGEMENT REVIEWS

Distribution:

1. Chairman
2. Director/General Manager
3. Factory Manager
4. Accountant
5. Factory Officer
6. Quality Officer
7. Administration Officer
8. Assistant Factory Officer – Withering
9. Assistant Factory Officer – Rolling
10. Assistant Factory Officer – Firing
11. Assistant Factory Officer – Sifting
12. Consultant – FSMS

1.0 Objective

The objectives of the procedure to review the food safety management system, to ensure its continuing suitability, adequacy, and effectiveness, assessing opportunities for improvements and changes in the food safety policy and food safety management system.

2.0 Scope

The scope covers all the activities related to the food safety management system of XYZ Tea Factory.

3.0 Resources

Meetings comprising of the following officers and shall be chaired by the General Manager.

Forum Members	Forum Quorum
1. Chairman 2. Director/General Manager 3. Factory Manager 4. Accountant 5. Factory Officer 6. Quality Officer 7. Administration Officer 8. Assistant Factory Officer – Withering 9. Assistant Factory Officer – Rolling 10. Assistant Factory Officer – Firing 11. Assistant Factory Officer – Sifting 12. Consultant – FSMS *Meeting Frequency – Quarterly*	07

4.0 Activities

4.1 Following subjects shall be discussed at the above meeting, which is considered as the inputs to the management review.

4.1.1 Follow-up actions from previous management reviews.

4.1.2 Analysis of results of verification activities of PRPs and hazard control plan.

4.1.3 Changing circumstances in external and internal issues that are relevant to FSMS, including changes in the company and its context that can affect food safety.

4.1.4 Emergency situations, accidents, and withdrawals.

4.1.5 Reviewing results of system-updating activities, performance information, and the effectiveness of FSMS, including trends in the result of system-updating activities, monitoring and measurement results, nonconformities, and corrective actions.

4.1.6 Review of internal and external communication activities, including customer feedback, requests, and complaints from interested parties.

4.1.7 Internal and external audit results with findings or inspections (regulatory, customer, any other interested parties) and performance of external providers.

4.1.8 Review of risks and opportunities identified in the process with the effectiveness of actions taken to address them and the extent to which objectives of the food safety management system have been met.

4.1.9 Opportunities for continual improvements.

4.2 The output from the management review shall include decisions and actions related to food safety.

4.2.1 Assurance of food safety with decisions and actions related to continual improvement opportunities.

4.2.2 Improvement of the effectiveness of the food safety management system.

4.2.3 If any need for updates and changes to the FSMS, including resource needs and revision of the food safety policy and objectives of the food safety management system.

4.2.4 Revision of the organization's food safety policy and related objectives.

4.3 Food safety management review meetings shall be held once every three months. The schedule shall be noticed one week in advance, and the minutes shall be circulated by the food safety team leader among the members.

5.3 The FST shall retain management review meeting input, output, and meeting minutes as evidence of the results of management review.

5.0 References

5.1 ISO 22000:2018 Standard

6.0 Records

6.1 Attendance Record – FSMS 04/FSP/10/F/01

6.2 Minutes of Management Review Meetings – FSMS 04/FSP/10/F/02

Name of Meeting/Training			
Date			
No	Name of Participant	Designation	Signature
1			
2			
3			
4			
5			
6			
7			
8			
9			
10			
11			
12			
13			
14			
15			
16			
17			
18			
19			
20			
21			
22			
23			
24			
25			
26			
27			
28			
29			
30			
31			
32			
33			
34			
35			
36			
37			
38			

ISO 22000 - PREREQUISITE PROGRAMME

PRP 01 - CONSTRUCTION, LAYOUT, AND EQUIPMENT

Distribution:

1. Director/General Manager
2. Factory Manager
3. Factory Officer
4. Quality Officer
5. Administration Officer
6. Assistant Factory Officer - Rolling
7. Assistant Factory Officer - Firing
8. Consultant - FSMS

1.0 Objective

To prevent and minimise food safety hazards being introduced to products due to layouts, interior design, equipment, and construction.

2.0 Scope

This procedure applies to buildings and equipment where the processing of food, storing of raw and packing material, in-process and finished products, including all amenities within the factory perimeter.

3.0 Responsibility

3.1 General Manager	– Guidance and compliance with requirements
3.2 Quality Officer	– Designing of layout and identification of appropriate equipment
	– Installation and maintenance of equipment
3.3 Factory Manager	– Identification of requirements
3.4 Consultant	– Compliance with FSMS
3.6 Factory Officer	– Monitoring and maintaining the facilities

4.0 Activities

4.1 Design and layout

4.1.1 The internal design and layout shall permit good food hygiene practices; prevent cross-contamination of finished made tea during and between operations from raw or in-process materials. When changes are made to the initial design and layout, those changes shall be carried out without deviating from the said requirements.

4.1.2 The internal design and layout shall permit adequate ventilation to prevent unacceptable accumulation of humidity condensation or dust and to remove contaminated air. Where applicable, all nets on the window are cleaned or washed once a month. Factory design shall ensure there is no risk of glass contamination in made tea.

4.1.3 Separate areas away from the production facility for vehicle parks should be identified and parked accordingly.

4.2 Internal Structures

4.2.1 Construction and finishing of the floor shall be so that will facilitate easy drainage and cleaning.

4.2.2 The construction material of the floor shall be resistant to the product manufactured in the specified area, detergents, and sterilising agents used.

4.2.3 Floors shall be sloped to avoid residual pools. The floors shall be sloped to drains or gutters at 1 inch 40 to 1 inch 60.

4.2.4 Sleeves or curbs shall be used as appropriate when pipes, ducts, or other equipment need to be passed through the floor.

4.2.5 Gutter drains shall have smooth vertical walls rounded or coved bottoms on a minimum slope of 1 in 100. Sizes are 152 to 304 mm wide and 152 to 203 mm deep. Gutters should be covered with a strong open metal grating rugged enough to withstand any unavoidable heavy traffic. These shall be in short sections of less than one meter so they can be easily lifted for cleaning. Grating shall be rested on ridges or buttons to allow water and small partials to drain away.

4.2.6 Wet areas shall have adequate and hygienic drainage systems. Floor drains shall be set in the floor so that the top of the grating is 3mm below the level of the floor. The floor drain shall have a minimum water seal of 36 mm. A removable secondary strainer shall be used to prevent the entry of cockroaches and rodents. Floors shall be slopped to drain.

4.2.7 Decking shall be made from solid plates or wood for stairs and platforms. Staircases that rise above the level of the product shall be encased to prevent dirt from being transferred into the product. Steps shall be made of the same anti-slip material.

4.2.8 Walls, partitions, doors, and windows, where appropriate, shall have smooth and non-absorbent surfaces, easy to clean, and where necessary, disinfect.

4.2.9 The paints suitable for incidental food contact or classed as food-grade shall be used on product zone surfaces. The paints that do not contain toxic or tainting materials shall be used in non-food zones.

4.2.10 Design for covers of wires, conduits, pipes, etc., where the system and the components shall be completely sealed or exposed so that they can be easily or thoroughly cleaned. Dust-tight or explosion-proof features shall be included in dry areas. Waterproofing and splash proofing shall be included in wet areas.

4.2.11 Ceilings and overhead fixtures shall be constructed and finished to minimise the accumulation of dirt and dust and to prevent condensation. Ceilings shall be non-porous easy to clean, and light coloured with a finish that has desirable light reflectance characteristics. The ceiling shall be free of the accumulation of dust and flaking paints. The junctures between walls and ceilings shall be rounded and sealed against dust and water. A service floor shall be used to house ducts, pipe works, cables as a floor foaming the ceiling of the production area to minimise overhead equipment and risk of contamination of the product.

4.2.12 Adequate natural or artificial lighting shall be provided such that the intended inspection or production activity can be effectively conducted. Where necessary, lighting shall

4.2.12.1 Be sufficient for the work or function.

4.2.12.2 Be of suitable colour where it is important, e.g., for inspection areas.

4.2.12.3 Provide suitable contrast between the work and the background.

4.2.12.4 Give minimum glair either from the light units or from the reflection from the other surfaces.

4.2.12.5 Give soft shadows if required but not deep shadowing.

4.2.12.6 Come from fixtures designed and installed to

- Avoid food/equipment contamination in the event of breakage.
- Be insects and dust-tight.
- Be easy to clean and maintain.

4.2.2 Recommended lighting levels in lux for each activity.

Location/ Operation	Minimum required light intensity (Lux) – (W/m2)
Dock /Loading and Unloading bays	300 – 500 (2.4 – 4.0 W/m2)
Receiving/ Withering area	300 – 500 (2.4 – 4.0 W/m2)
Rolling Room	350 – 500 (2.8 – 4.0 W/m2)
Firing	350 – 500 (2.8 – 4.0 W/m2)
Sifting/ Blending/Packing	400 – 500 (3.2 – 4.0 W/m2)

4.2.2.1 Emergency lighting - Emergency lighting is required in areas where a normal lighting system may fail.

4.2.2.2 Escape lighting – This both shows and lights escape roots and enables orderly evacuation as well as assisting emergency services by locating fire fighting and first aid equipment.

4.2.3 Condition of all internal structures and fittings shall be inspected monthly by the process owner and /or food safety team member and shall be informed to the administrative officer for corrective actions, if any.

4.3 Preventing access (exclusion measures) for Rodents, Insects, and Birds.

4.3.1 Outer boundaries of factory areas, the walls around the buildings shall be surrounded by tarmac or well-paved path preferably 5-6m wide, which must be kept as clean and empty as possible.

4.3.2 When there is no tarmac path, the growth of weeds shall be prevented by carefully scheduled use of weed killer, and the ground shall be kept flat and levelled as possible. The growth of vegetation shall be controlled effectively by using gravel to cover the ground or utilizing the land as appropriate.

4.3.3 Drainage channels and gratings shall be kept empty and clean. Drainage gratings and corners shall be easy to remove for routine maintenance, but at the same time, mesh grills must be firmly set in place with drainage holes designed to prevent the access of rodents.

4.3.4 The parking of pallets, equipment, and goods in this area must be kept short. This area shall be well-drained.

4.3.5 Foundations shall be in concrete and go down to the minimum of 300mm (1feet) below solid ground level.

4.3.6 In the case of potentially gnawable walls, a smooth metal sheet at least 300 mm wide (1 ft) shall be used to cover the base of the walls.

4.3.7 Provisions shall be made to arrange special cleaning regimes for the architectural stone pillars on the factory.

4.3.8 The entry doors to production facilities shall be equipped with automatic self-closing devices.

4.3.9 Openings around the pipes leading to and away from the premises must be protected, e.g., sealed with wire netting, which is then covered and filled with cement.

4.3.10 Windows shall be protected with anti-insect netting, preferably using a virgin plastic mesh of 2mm.

4.3.11 Air intakes/vents shall be equipped with anti-insect netting. If the vents are located in critical production areas, they must be equipped with special (HEPA) filters.

4.3.12 Lighting in operational areas shall be such to permit very accurate work and also efficient inspection. Positioning of lights directly over areas where the products are exposed should be avoided.

4.3.13 In places where doorways directly link to the production department from an outdoor environment, a chamber with a double door shall be used. The door shall be equipped with automatic door closures. This chamber shall be maintained dark, and an insecticuter shall be in operation.

4.3.14 To prevent access of birds to the building as a whole, shall be free of window sill's ledges and protuberances as possible, thus not giving birds a chance to lodge or nest.

4.3.15 There shall be a designated area for storage of waste and segregated non-conforming products, which shall prevent cross-contamination with made tea from waste/non-conforming products.

4.4 Equipment

4.4.1 Equipment and containers which are directly in contact with food shall be designed and constructed to ensure that they can be easily and adequately cleaned, disinfected, and maintained to prevent contamination of food.

4.4.2 The material used for equipment and containers shall be of nontoxic nature for the intended use. A list of materials approved by the food safety team leader for use as food contact surfaces shall be maintained in each production unit.

5.0 References

5.1 Plant layout

6.0 Records

6.1	Suitability of internal structures and fittings	FSMS 04/PRP/01/F/00
6.1.1	Receiving/ Withering area	FSMS 04/PRP/01/F/01
6.1.2	Rolling room	FSMS 04/PRP/01/F/02
6.1.3	Firing	FSMS 04/PRP/01/F/03
6.1.4	Sifting & Packing	FSMS 04/PRP/01/F/04
6.1.5	Packing material stores	FSMS 04/PRP/01/F/05
6.1.6	Finished product stores	FSMS 04/PRP/01/F/06
6.1.7	General stores	FSMS 04/PRP/01/F/07
6.1.8	Materials approved as food contact surfaces	FSMS 04/PRP/01/F/08

Department				Date Inspected
D. Ref FSMS 04/PRP/01/F/01	Suitability		Name of the Auditor	
Internal structure/Fitting	Yes	No	Comments	Recommendation

ISO 22000 - PREREQUISITE PROGRAMME

PRP 02 - MANAGEMENT OF PURCHASED MATERIALS

Distribution:

1. Director/General Manager
2. Factory Manager
3. Factory Officer
4. Quality Officer
5. Administration Officer
6. Assistant Factory Officer – Withering

1.0 Objective

To ensure the purchase of good quality and safe green leaf and other materials which conform to specified requirements, on time for the manufacturing processes of the company.

2.0 Scope

This will cover the registering, requesting, ordering, and purchasing of raw materials requires for the production of orthodox black tea manufacture and packaging materials and any others that affect food safety or quality, including local purchases and imports.

3.0 Responsibility

3.1	General Manager	- Guidance and compliance with factory requirements
3.2	Quality Officer	- Determination & conformation of suitability
		- Identification of requirements
3.6	Factory Officer	- Supplier evaluation
3.7	Administrative Officer	- Sourcing of suppliers and purchasing
		- Communicating the factory requirements
3.5	Assistant Factory Officers	- Raw material inspection and verification

4.0 Activities

4.1 All raw and packing materials shall be purchased from approved, registered suppliers, and all such suppliers shall be subjected to regular evaluation. (Process Flow Diagram for supplier evaluation, FSMS 04/PRP/12/F1).

4.2 The raw materials shall have clearly defined green leaf plucking and preparation specifications according to the TRI/SLTB, and the plucked leaf shall be assessed, weighed and origin recorded at the point of collection. Thus, purchasing of raw material shall be done according to the published industry norms of brought leaf for black tea manufacture by Tea Research Institute, and the standard method of "Ran Dhalu " (Green leaf verification) shall be applied at the field as well as at factory receiving for production.

4.3 The raw materials shall be assessed for contamination of foreign matter. The suppliers shall be monitored and controlled for the use of pesticides and other potential chemical contaminants through communicated quality and consumer safety standards.

4.4 Packing materials shall be free from risks of potential contamination of made tea, and all the packing materials shall be traceable back to the original manufacturer/ supplier. The factory shall use a standard approval process for all packing materials, where all packing materials that come into direct contact with made tea shall conform to standard food safety specifications. The packing materials shall have printed information as per the standard food safety requirements.

4.5 When a new material or a new supplier is to be introduced, samples shall be requested and checked by the factory manager to ensure suitability. The quotations shall be called. Samples shall also be called together with the quotation as and when required.

4.6 For other purchases of supplies, the material shall be evaluated according to the material specification and safety/quality parameters of the specific material. Once the standard approval is granted, Purchase Order (PO) shall be generated and conveyed to the supplier with appropriate lead time. A certificate of analysis shall be requested as per the list of materials that require a certificate of analysis.

4.7 Materials brought into the stores shall be examined by Clerk/Forman before acceptance for correct material, damages, evidence of pest infestations, and label requirements (WI/ST/01 – Raw & Packing Material Handling).

4.8 The out-of-specification green leaf or if any item is not complying with the predefined standards, such material shall be rejected. (Information on such rejections shall be documented)

4.9 Notification of rejection shall be conveyed to the supplier through purchasing department.

5.0 References

5.1 Supplier Evaluation & Selection – FSMS 02/PRP/12

5.2 Testing schedules

5.3 Specifications for incoming raw and packaging materials

5.4 Testing methods

5.5 Work Instructions

5.6 Process flow chart for supplier evaluation - FSMS 04/PRP/12/PFD/01

6.0 Records

6.1 Audit checklist for Supplier Evaluations

6.2 Green leaf quality inspection record - FSMS 04/PRP/12/RC/01

6.3 "Ran Dhalu" verification at receiving (Leaf Count Book) – FSMS 04/PRP/02/RC/01

6.4 Certificate of analysis for packaging and other materials

6.5 Packing material receiving & inspection record – FSMS 04/PRP/02/ RC/02

PACKING MATERIAL RECEIVING AND INSPECTION RECORD

Date	Selling Mark	Quantity	Supplier		Material Type	Packing Material		Delivery Vehicle		Responsibility (Purchasing Officer)	Remarks	FSTL's Comments (Verification)
			Name	Rank		Quality	Hygiene	Condition	Hygiene			

ISO 22000 - PREREQUISITE PROGRAMME

PRP 03 – PREVENTIVE MAINTENANCE

Distribution:

1. Director/General Manager
2. Factory Manager
3. Factory Officer
4. Assistant Factory Officers
5. Consultant – FSMS

1.0 Objective

The objective of this PRP is to ensure the suitability of plants and machinery for the production of quality and safe products.

2.0 Scope

This will cover all buildings, plants, machinery, equipment, utensils, tools, and vehicles used for the processing of orthodox black tea.

3.0 Responsibility

3.1	General Manager	– Guidance & compliance with factory requirements
3.2	Quality Officer	– Identification of requirements & implementation
		– Planning & implementation
3.3	Factory Officer	– Scheduling
3.4	Assistant Factory Officers	– Monitoring the effectiveness

4.0 Activities

4.2 A preventive maintenance plan shall be prepared based on manufactures recommendations, history records, and criticality to food safety.

4.3 Schedule of cleaning and maintenance program for machinery and equipment shall be carried out.

4.4 All equipment and machinery belong to a specified area should be identified, documented, and maintained.

4.5 The processing equipment shall be designed to facilitate effective and efficient use while minimizing the product contamination, where equipment shall be made of food-grade material with smooth, impervious, non-toxic, non-absorbent, and corrosion-resistant material in all the product surfaces that are directly in contact with products to reduce the contamination.

4.6 The equipment and utensils or machinery shall be free of rough surfaces or cracked walls to prevent contamination and accumulation of materials, where equipment shall be constructed to preclude metal to metal contact.

4.7 The equipment and utensils shall be designed to preclude or divert condensations away from product and product contact surfaces.

4.8 The machinery, equipment, or utensils shall have appropriate lids/covers to protect the product from unintended contamination and accidents.

4.9 Valves, gaskets, and pumps of the equipment shall be maintained in a clean and sanitary manner, where all processing equipment shall be maintained free of leaks at valves, gaskets, fittings, etc. and non-food grade materials such as wire, tape, string, wood, and cardboard shall not be used as a temporary design or repair.

4.10 The machinery, equipment, and utensils shall be free of oil leaks, excessive grease build-ups, or bearings in the motor housing, and also free of flaking paint and rust to reduce risks of contamination.

4.11 The processing equipment shall be installed to provide easy access for cleaning and inspections as specified. The conveyor belts shall be free of loose string, metal clips, and loose pieces of rubber.

4.12 Breakdown maintenance shall be carried out as and when required. Protective garments shall be worn by the maintenance staff during maintenance work.

4.13 Adequate precautions shall be taken to avoid contamination of the work environment during maintenance work. The food-contact surfaces, processing equipment, control panels, gauges, and instruments in the vicinity shall be protected when surface preparation and painting of buildings or equipment is carried out. Ingredients, products, or packaging materials shall be protected from materials used for painting such as thinner, which can cause toxicity, taint problems, or absorption of odour.

4.14 Repaired plant or machinery shall be adequately cleaned and disinfected before the commencement of production as per the instructions of the assistant factory officer after each breakdown in the case of production machinery.

4.15 Process capability tests shall be carried out if and when processes and equipment are known to be incapable or equipment has undergone a repair or changes which requires re-assessment of capability.

4.16 Maintenance of water treatment unit, air compressors, oven, sieves, and colour sorters shall be carried out in such a way that it will not contaminate products.

4.17 Measuring equipment shall be controlled and calibrated.

5.0 References

5.1. FSP for corrective action and correction – FSMS 02/FSP/06

5.2. Work instructions/service manuals for service equipment.

5.3. Process flow chart for control of monitoring and measuring equipment.

6.0 Records

6.0 List of plant and equipment

6.1 Equipment history record

6.2 Preventive maintenance plan

6.3 Preventive maintenance checklist – FSMS 04/PRP/03/CL/01

6.4 Preventive maintenance record – FSMS 04/PRP/03/RC/02

Preventive Maintenance Check List						
Plant Name				**Department**		
D. Ref	FSMS 04/PRP/03/CL/01		**Responsible Officer**			
Equipment Name	Date Serviced	Next Service	Part Replaced	Replacements Required	Technician	Remarks

D. Ref: FSMS 04/PRP/03/RC/01

Preventive Maintenance Record								
Name of Machine/ Equipment			Dept:				Code No.	
Date		Running		Replacements			Responsibility	FSTL's
No	Inspection Criteria	Status	Condition	Completed	Required	Planned	Technician	Comments
1								
2								
3								
4								
5								
6								
7								
8								
9								
10								
11								
12								
13								
14								
15								
16								

ISO 22000 - PREREQUISITE PROGRAMME

PRP 04 - CONTROL OF MONITORING, MEASURING EQUIPMENT, AND METHODS

Distribution:

1. Director/General Manager
2. Factory Manager
3. Factory Officer
4. Administration Officer
5. Assistant Factory Officer – Rolling
6. Assistant Factory Officer – Firing

1.0 Objective

The objective of the PRP is to ensure that the results obtained from monitoring and measuring activities are accurate and reliable.

2.0 Scope

The scope covers all equipment and methods used for monitoring and measuring activities.

3.0 Responsibility

3.1	General Manager	– Oversee the operation
3.2	Quality Officer	– Ensure proper methods & equipments are in place Ensure all equipments are properly maintained & calibrated at planned intervals
3.3	Factory Officer	– Ensure all equipments are properly maintained & calibrated at planned intervals
3.4	A F O – Firing	– Conducting of internal calibration & external calibration of monitoring & measuring equipments

4.0 Activities

4.1 All equipment used for monitoring and measuring shall be controlled. All inspection, measuring, and test equipment, which are used in all phases of product/process verification, are controlled and calibrated against nationally traceable standards at specified intervals.

4.2 All equipment/instruments in the calibration system shall be identified with stickers that indicate the calibration status and calibration certificates are maintained, where applicable.

4.3 All measuring, test equipment, controlling production/servicing processes shall be calibrated at prescribed intervals. The calibration system further extends to manufacturing orthodox black tea and process instrumentation that can affect product quality or is used to measure specified characteristics.

4.4 Equipment that is used for reference only (not verification) is not included in the calibration system, and that is labelled with "DO NOT USE FOR VERIFICATION" stickers warning that it is not calibrated.

4.5 All employee-owned measuring instruments used for verification of products are registered with quality assurance and subject to all controls of the calibration system. Employees shall not, under any circumstances, use their measurement instruments if they are not registered and calibrated. If any controlled equipment fails a routine performance check, the Quality Officer shall be informed immediately.

4.6 Appropriate inspection, measuring, and test equipment shall be selected to satisfy the accuracy and precision of the required characteristics. The items shall be purchased from reliable suppliers capable of verifying that the item meets the requirements.

4.7 A Controlled Equipment Calibration Master Schedule shall be maintained by the QO. All active equipment shall be entered on a controlled list, indicating the calibration interval. All inspection, measuring, and test equipment used for verification purposes are calibrated to nationally or internationally traceable standards and shall be labelled with a calibration sticker. Where practical, the instruments shall be sealed with tamper-proof seals to prevent unauthorized repair or adjustment.

4.8 Calibration shall be performed either by external calibration laboratories or internally. When external services are utilized, they shall be incorporated into the Approved Supplier List and are monitored via the company's supplier rating system to prevent the use of labs that fail to meet quality and performance requirements.

4.9 All internal calibration shall be performed in accordance with written instructions.

4.10 Equipment that is calibrated internally according to the internal calibration procedure and records are maintained in the internal calibration logbook. The Calibration Test report shall be stored for one year from the date of calibration.

4.9 All equipment shall be calibrated in a suitable environment and is handled with care to assure damage is not sustained, and the calibration is not affected.

4.10 Certificates of calibration shall be maintained on file for all instruments that were calibrated by accredited external calibration laboratories.

4.11 Each instrument shall be traceable to its own Calibration History Record, which contains its identification number, storage location, make or type, frequency of calibration, reference standards used, actual calibration findings including date, and actions to be taken in case of unsatisfactory results.

4.12 The QO shall be responsible for extending the control of measuring and test equipment to all approved suppliers.

4.13 Nonconforming Equipment

4.13.1 If it is confirmed that the equipment is out of calibration, the QO shall investigate the validity of measurements for which the equipment was previously used and assesses the acceptance status of the entire affected product.

4.13.2 If applicable, all nonconforming products shall be identified and dispositioned as described in the FSMS 02/FSP/07. In the event that a nonconforming product has been entered into the market, the customer shall be immediately notified.

4.13.3 Any measuring or test equipment that appears to give inaccurate readings is checked and calibrated.

4.14 Monitoring and measuring shall be carried out as per the documented test methods and monitoring procedures.

4.15 The test method shall be verified annually through external laboratories.

4.16 The usage of measuring and monitoring equipment is restricted, and the measuring instruments shall be adjusted only by the authorized personnel.

4.17 The Standard Operating Procedures shall be displayed at the equipment or nearby and communicated to the authorized personnel who will carry out operations.

4.18 The calibration status shall be displayed on the equipment.

4.19 The records of calibration and verification results shall be maintained.

5.0 References

5.1 Test Methods

5.2 Calibration Records

5.3 Calibration Master Schedule

6.0 Records

6.0 List of monitoring & measuring equipment at the factory.

6.1 List of monitoring & measuring equipment at the field.

6.2 Details of monitoring & measuring equipment for internal/external calibration. (FSMS 04/PRP/04/F1)

6.3 Internal calibration reports. (FSMS 04/PRP/04/F2)

6.4 External calibration reports. (FSMS 04/PRP/04/F3)

ISO 22000 - PREREQUISITE PROGRAMME

PRP 05 – STORING, HANDLING, AND TRANSPORTATION OF MATERIAL

Distribution:

1. Director/General Manager
2. Factory Manager
3. Factory Officer
4. Quality Officer
5. Assistant Factory Officers
6. Consultant – FSMS

1.0 Objective

The objective of this PRP is to maintain and manage the conditions to prevent contamination and to provide adequate facilities for storage, handling, and transportation of food and non-food material.

2.0 Scope

This PRP covers all raw, in-process, packaging material, finish products, and non-food materials or chemicals.

3.0 Responsibility

3.1	General Manager	– Providing Guidance and establishing requirements – Ensuring required resources are available
3.3	Factory Manager	– Obtaining required resources and proper handling & storage – Ensuring compliance to FSMS requirement at suppliers' points
3.2	Quality Officer	– Ensuring all requirements are implemented
3.4	Assistant Factory Officers	– Ensure proper handling & storage
3.5	Consultant – FSMS	– Ensuring the adherence to FSMS

4.0 Activities

4.1 Storing

4.1.1 Stipulated conditions shall be provided and maintained to prevent deterioration of material quality and safety as per work instruction for handling and storage.

4.1.2 Storing shall be done in predefined locations that will ensure adequate space from walls and stored material, floor and stored material that permit adequate maintenance and cleaning to avoid pest access and harbourage.

4.1.3 All non-food chemicals (detergents) shall be stored separately from all food materials and packing materials to prevent cross-contamination.

4.1.4 Any toxic compounds allowed in the plant shall be identified, held, used, and stored in a manner that protects against contamination of food, food-contact surfaces, or packaging materials.

4.1.5 Issuing of goods shall be done according to the FIFO system.

4.1.6 Containers holding half-used materials shall be kept closed to prevent cross-contamination.

4.1.7 Batch identification shall be maintained in all raw, in-process, packing material, and in finished products.

4.1.8 All storage areas shall be cleaned as per the cleaning schedule and maintained to ensure a proper hygienic environment. Records shall be maintained.

4.1.9 Stock levels shall be maintained according to the Re-order level.

4.2 Handling

4.2.1 Necessary care shall be taken to prevent damages, spillages, and spoilages in raw, in-process, and finished products.

4.2.2 Stipulated conditions shall be provided and maintained to prevent deterioration of material quality and safety during handling. Green leaf storage conditions (withering) shall be provided with care, and measurements of temperature shall be taken as to the standards while maintaining adequate records (FSMS 04/PRP/05/RC/01 - 02).

4.2.3 Special attention shall be given to prevent cross-contamination.

4.2.4 All production and store staff shall be trained on the handling of relevant material. Relevant Work Instructions shall be displayed.

4.3 Transportation

4.3.1 Necessary care shall be taken to prevent damages, spillages, and spoilages while transporting materials, especially in brought leaf transportation.

- Use crates for transport of green leaves and weights up to 25Kg total, including create.
- Stack creates up to standard heights.
- Reduce transportation time as much as possible.
- Poly/gunny sacks shall be filled up to 22Kg, including the sack's weight.
- Visual observation while loading or filling crates for material food safety aspects and quality aspects as to the company specifications.

4.3.2 Vehicles use for transportation shall be maintained with appropriate conditions to prevent product contamination and physical damages.

4.3.3 The cleanliness of the finished product transport truck shall be inspected and recorded by the storekeeper prior to loading.

4.3.4 Raw material transportation trucks and bulk transportation containers shall be inspected for cleanliness and suitability at predefined intervals. Records shall be maintained.

4.3.5 Made black tea shall not be exposed to adverse weather conditions during transportation.

5.0 References

5.1. Correction & Corrective Action - FSMS 02/FSP/06

5.2. Work Instruction for handling and storage

5.3. Work instructions for brought leaf transport crew

5.4. Work instructions for transportation of made tea

6.0 Records

6.1	Checklist for cleaning, hygiene and suitability of crates, bulk containers	- FSMS 04/PRP/05/CL/01
6.2	Vehicle inspection checklist	- FSMS 04/PRP/05/CL/02
6.3	Hygrometer record	- FSMS 04/PRP/05/RC/01
6.4	Withering record	- FSMS 04/PRP/05/RC/02

Cleaning Check List for Bulk Containers/Crates							
Plant Name					**Department**		
D. Ref	FSMS 04/PRP/05/CL/01		**Responsible Officer**				
Container	Application	Point of Use	Date Inspected	Hygienic Condition	Cleaning Quality	Suitability	Remarks

Cleaning Check List for vehicles									
Plant Name							Department		
D. Ref	FSMS 04/PRP/05/CL/02		Responsible Officer						
Vehicle No	Application	Point of Use	Date Inspected	Hygienic Condition			Checked By	Remarks	Verified By
				Cleaning	Body/hood	Overall			

Black Tea

XYZ TEA FACTORY
ISO 22000 - PREREQUISITE PROGRAMME
WITHERING RECORD

D. Ref: FSMS 04/PRP/05/RC/02

HYGROMETER TEMPERATURE RECORD

Time of the Day (Temperatures Measured in °F)

Date & Tough No	Bulb Type	04.00 pm	05.00 pm	06.00 pm	07.00 pm	08.00 pm	09.00 pm	10.00 pm	11.00 pm	12.00 pm	01.00 am	02.00 am	03.00 am	04.00 am	05.00 am	06.00 am	07.00 am	08.00 am	09.00 am	10.00 am	11.00 am	12.00 am	01.00 pm	02.00 pm	03.00 pm
	Dry																								
	Wet																								
	Diff																								
Checked By																									
Reviewed By																									
	Dry																								
	Wet																								
	Diff																								
Checked By																									
Reviewed By																									
	Dry																								
	Wet																								
	Diff																								
Checked By																									
Reviewed By																									
	Dry																								
	Wet																								
	Diff																								
Checked By																									
Reviewed By																									
	Dry																								
	Wet																								
	Diff																								
Checked By																									
Reviewed By																									
	Dry																								
	Wet																								
	Diff																								
Checked By																									
Reviewed By																									

Prepared : FSTL
Reviewed : C - FSMS
Approved : D/GM

Revision : 00
Date : 01-11-2018
Page : Page 1 of 1

Black Tea

XYZ TEA FACTORY
ISO 22000 - PREREQUISITE PROGRAMME
WITHERING RECORD

D. Ref: FSMS 04/PRP/05/RC/03

Trough No	Date	Start	Withering Time						End	Supplier/ Line Code	Release Time/Code	Responsibility (A.F.O - WI)	FSTL's Comments (Verification)
			1st Mixing		Turning		2nd Mixing						
			Start	End	Start	End	Start	End		Vehicle No			

Prepared : FSTL
Reviewed : C - FSMS
Approved : D/GM

Revision : 00
Date : 01-11-2018
Page : Page 1 of 1

VEHICLE INSPECTION RECORD

Date	Receiving Time	Vehicle No	Supplier Details			Condition of Vehicle			Driver	Checked By	Remarks	FSTL's Comments (Verification)
			Supplier Code	Area	Qty (Kg)	Cleaning	Body/hood	Overall				

DISPATCHED RECORD

Date	Dispatched Time	Vehicle No	Dispatched Details			Condition of Vehicle			Driver	Checked By	Remarks	FSTL's Comments (Verification)
			Batch Code	Dis. Qty (Kg)	No of Bags	Cleaning	Body/hood	Overall				

ISO 22000 - PREREQUISITE PROGRAMME

PRP 06 – PERSONAL HYGIENE

Distribution:

1. Director/General Manager
2. Factory Manager
3. Factory Officer
4. Quality Officer
5. Assistant Factory Officer – Withering
6. Assistant Factory Officer – Rolling
7. Assistant Factory Officer – Firing
8. Assistant Factory Officer – Sifting
9. Consultant – FSMS

1.0 Objective

The intention of this PRP is to prevent contamination from personnel engaged in product-related operations and visitors to the factory.

2.0 Scope

This procedure covers health status, cleanliness, the behaviour of employees engaged in the handling of foods and visitors.

3.0 Responsibility

3.1	General Manager	– Guidance & compliance with factory requirements
3.3	Factory Manager	– Providing resources
		– Providing necessary training
		– Protective uniforms
3.2	Quality Officer	– Implementation & coordination
3.4	Assistant Factory Officers	– Implementation monitoring the effectiveness
3.5	Consultant – FSMS	– Providing guidance
		– Establishing requirements

4.0 Activities

4.2 Annual medical check-ups shall be carried out to ensure the fitness for engagement of work. Following tests shall be carried out.

- Nasal swab for *Staphylococcus aureus*
- Faecal test/ Stool for *Salmonella & Shigella*
- Faecal/ Stools full report
- Full blood count
- ECG (If necessary)
- X-ray (If necessary)

4.3 All reports shall be reviewed by the company doctor, and medicine shall be prescribed if required. Employees found positive shall be allocated to work points that do not allow direct contact with food material until they are tested and confirmed cured after medication.

4.4 Any employee who has or may have, by medical examination or supervisory observation, an illness, infected wound, an open lesion such as a boil or sore, or any other problem that might contaminate food, food-contact surfaces, or packaging materials shall be excluded from any operations until the condition is healed or corrected.

4.5 The employees known, suspected to be suffering from contagious diseases (Jaundice, diarrhea, vomiting, fever, and sore throat with fever, visibly infected skin lesions, discharges from the ear, eye, or nose) shall be immediately vacated from the production areas. Records shall be maintained.

4.6 Employees shall be made aware of reporting to the management if they are suffering from any disease or symptom mentioned in 4.4. (As per work instructions). Employees shall notify immediate supervisors if they have been exposed to a confirmed disease outbreak of Salmonella (such as typhoid), hepatitis A or *Shigella*, especially when employees are asymptomatic.

4.6.1 Hand washing; shall be applied as both operative and management requirement. Hands shall be washed before leaving the toilet block and each time while entering food-handling areas. Within the area, hands shall be washed when they get dirty.

4.6.2 Eating, drinking, and smoking shall be prohibited in food processing areas.

4.7 Food handlers shall maintain a high degree of personal cleanliness and wear suitable protective clothing, head covering, and footwear as per the protective garment guide for each department. Each food handler shall be provided with a clean uniform daily, and the used/ dirty uniforms shall be drop-in dirty clothes bins that shall be rewashed by them after the work or laundry services can be used. All food handlers shall wash their hands whenever they engage in food processing.

4.8 Cotton reusable/disposable gloves shall be provided for food handlers to maintain food safety. The gloves shall be washed or discarded when the wearer leaves the work area, and a new pair shall always be worn by the food handlers when the work is recommenced.

4.9 Personnel having cuts and wounds shall be permitted to continue working with suitable waterproof dressings. Such incidents shall be recorded in the personnel hygiene register of each production/storage unit.

4.10 Work instructions on personnel hygiene shall be given to factory employees where necessary

4.11 The level of personal hygiene of individuals engaged in food handling shall be monitored weekly and recorded in the personnel hygiene inspection record book.

4.12 Refresher training shall be provided to all employees on personal hygiene annually.

4.13 The visitors shall be dressed in necessary protective garments when entering to specified areas on the premises.

4.14 The engineering staff and the maintenance contractors shall be dressed in clean uniforms/overalls and changed regularly.

4.15 Smoking, chewing beetle, and chewing gums shall be permitted only in the specified locations of the factory premises.

5.0 References

5.1 Personal hygiene general instructions
5.2 Health control guide – FSMS 04/PRP/06/GI/01

6.0 Records

6.1 Reports on annual medical check-ups – FSMS 04/PRP/06/RP/01
6.2 Personnel hygiene inspection record book – FSMS 04/PRP/06/RB/01
6.3 Protective garment register – FSMS 04/PRP/06/RG/01
6.4 Personnel illness and injuries register – FSMS 04/PRP/06/RG/02

User Guide:
Each employee, including the staff, shall be checked for the following criteria twice a week before entering to the production facility, which is critical to food safety and to improve the prevention of cross-contamination. If the requirements are satisfactory, mark with "√" in the relevant cage; if not, isolate the employee and rectify the issue or act as appropriate.

Health Check Record Book																			
Plant Name							**Responsible Officer**												
Verification		Employee EPF Number																	
	Aspect																		
1	Trimmed hair																		
2	Trimmed nails																		
3	Trimmed moustache																		
4	Compliance with ornament policy																		
	Jewellery, chain, bangles, earring																		
	Wedding ring																		
5	Wearing proper uniform																		
	Hair cap																		
	Mask																		
	Apron																		
	Slippers																		
6	Hand washing																		
7	Reporting of illnesses																		
8	Visually observed illnesses																		
9	Wounds and prevention steps																		
10	Any inappropriate behaviors																		
	Smoking																		
	Betel Chewing																		
	Use of perfumes/aftershaves																		

Protective Garment Supply Register

Plant Name				Department				
Responsible Officer								
Date	Employee Name	Uniform	Apron/ Over Coat	Cap	Mask	Gloves	Boots	Goggles

Name of Employee:			
Designation:		**ETF No.**	
Section:			
Nature of the Duty:			
Age:		**Sex:**	

Test Criteria

The intention of this activity is to prevent contamination from personnel engaged in product-related operations.

Annual medical check-ups shall be carried out to ensure the fitness for engagement of work.

Following tests shall be carried out and shall attach all test reports to the individual document.
Nasal swab for Staphylococcus aureus
Faecal test/ Stool for Salmonella & Shigella (Faecal/ Stools full report)
Full blood count
ECG (Optional – Decided by the medical practitioner)
X-ray (Optional – Decided by the medical practitioner)

General Check-up
CNS
ENT
Cardio Vascular System (Heart Rate, Pressure, Auscultation)
Other (any other tests as decided by the medical practitioner)

Test 01	**Nasal swab for Staphylococcus aureus**				
Doctor's Comments					
Special Remarks	**Report Attached**	**Yes**		**No**	
If any:					

Test 02	**Faecal test/ Stool for Salmonella & Shigella (Faecal/ Stools full report)**				
Doctor's Comments					
Special Remarks	**Report Attached**	**Yes**		**No**	
If any:					

Test 03	Full blood count				
Doctor's Comments					
Special Remarks	Report Attached	Yes		No	
If any:					

Test 04	ECG				
Doctor's Comments					
Special Remarks	Report Attached	Yes		No	
If any:					

Test 05	X ray				
Doctor's Comments					
Special Remarks	Report Attached	Yes		No	
If any:					

<table>
<tr><td colspan="2">Test 06</td><td colspan="5">General check-up (CNS/Pressure/ENT/Other)</td></tr>
<tr><td colspan="7">Doctor's Comments</td></tr>
<tr><td>CNS</td><td colspan="2">UFR</td><td>ENT</td><td colspan="2">CVS</td><td>Other</td></tr>
<tr><td></td><td colspan="2"></td><td></td><td colspan="2"></td><td></td></tr>
<tr><td>Special Remarks</td><td>Reports Attached</td><td>Yes</td><td></td><td>No</td><td colspan="2"></td></tr>
<tr><td colspan="7">If any:</td></tr>
</table>

*CVS = Cardio Vascular System = Heart Rate + Pressure + Auscultation

Recommendation of the Medical Officer:

SCHEDULE
(Regulation 11)

I, Dr. ……………………………… Medical Officer of (Institute)…………………………………………… have physically examined after appropriate laboratory investigations on Mr./Mrs…………………… ……………………of XYZ Tea Factory Limited, certify that he/she is not suffering from any communicable disease or carrier of wounds, skin infections, sores, or diarrhoea.

……………………	……………………	……………………
Name:	Signature:	Date:
(Official Stamp)		

Illness and Injuries Register

Plant Name		Department	
Responsible Officer			

Date	Name/ EPF No	Reason (illness or injury)	Informed to	Work point	New work point	Dates removed from the working point

ISO 22000 - PREREQUISITE PROGRAMME

PRP 07 – PEST CONTROL

Distribution:

1. Director/General Manager
2. Factory Manager
3. Factory Officer
4. Quality Officer
5. Assistant Factory Officer – Withering
6. Assistant Factory Officer – Rolling
7. Assistant Factory Officer – Firing
8. Assistant Factory Officer – Sifting
9. Consultant – FSMS

1.0 Objective

The objective of this PRP is to ensure the factory premises free from pests and to prevent the likely hood of infestation from the incoming materials.

2.0 Scope

The scope covers all production facilities, stores, areas of supportive services, and the open area inside the premises.

3.0 Responsibility

3.1	General Manager	– Guidance & compliance with factory requirements
3.2	Factory Manager	– Ensuring implementation & monitoring contractor
3.3	Quality Officer	– Identification of requirements & communicate to GM
		– communicating with the contractor
3.4	Asst. Factory Officers	– Identification of requirements & communicate to GM
3.5	Consultant – FSMS	– Providing guidance and establishing requirements
		– Ensuring the FSMS requirements are fulfilled

4.0 Activities

4.2 All buildings shall be kept in good repair and condition to prevent pest access and eliminate potential breeding sites (PRP 01 Construction and layout).

4.3 Measures shall be taken to prevent/control any animals entering to the factory premises. The perimeter wall and the fence shall be properly maintained to prevent the entrance of animals.

4.4 Pest sighting incidents shall be reported, recorded, and conveyed to the pest control technician in every visit, or if there is an emergency pest infestation, immediately contact the control technician.

4.5 Prevention of harbourage and infestation/Internal hygiene and cleanliness of premises.

4.5.1 Production premises shall be kept extremely tidy. Various types of materials shall not be allowed to accumulate, and product leftovers and rejects must not be left standing in the area (refer PRP – 11).

4.5.2 Unused space shall be kept scrupulously clean to reduce hiding and nesting places for insects. E.g., corners and spaces behind and below machinery, between machinery.

4.5.3 Rework shall be collected in properly designed separate containers/bins.

4.5.4 Walls and floors shall be kept clean, and care will be taken to ensure that there are no cracks, crevices, or holes.

4.5.5 Drainage channels shall be kept clean and closed to prevent pest harbouring.

4.5.6 Production machinery and equipment shall be set above the ground or sealed to the plinth or floor to eliminate infestation locations.

4.5.7 All windows shall be kept close or netted, and all ventilation ducts shall be covered with meshes to avoid pests coming in.

4.5.8 All doors shall be kept closed with door closures, and air/strip curtains will be installed where appropriate. A door from a production facility shall never be opened directly to outside; hence double doors will be fixed where appropriate.

4.5.9 All internal and external structures shall be designed and constructed to facilitate adequate drainage and cleaning to prevent harbourage and infestation of pests. (PRP 01– Construction, Layout, and Equipment)

4.5.10 All materials accepted into company warehouses shall be free from evidence of pest infestations. (PRP 02 – Management of Purchased Materials)

4.5.11 Food items shall be kept in pest-proof containers and /or stacked above the ground and away from the walls.

4.5.12 All refused items where appropriate shall be kept in pest-proof containers. Waste material stores shall be kept clean to avoid harbourage and infestation.

4.6 Monitoring & Detection

4.6.1 Factory premises shall be examined for evidence of infestation by an Assistant Factory Officer or a Food safety member (Checklist for the pest control programme, FSMS 04/PRP/07/CL/01). The pest control technician shall go through the checklist before carrying out the treatments.

4.6.2 Positive evidence of infestation shall be communicated immediately to the Consultant – FSMS via email for corrective action.

4.7 Eradication

4.7.1 Pest infestations shall be dealt with immediately and without adversely affecting the suitability and safety of food. Chemicals approved by the pesticide registrar shall only be used for eradication.

4.7.2 All chemicals and equipment used in pest control shall be stored under lock and key in a separate location away from food processing areas.

4.7.3 Approved chemicals for food production companies shall be used for all pest control activities (Refer pest treatment plan for chemicals used).

4.7.4 The bait stations shall be covered boxes or structure which shall be designed to minimize or eliminate accidental contamination or migration to the food or packing materials. Hence standard bait stations shall be used inside and outside the facility; rat traps without pesticides are recommended inside.

4.7.5 All pest control treatments are done by a skilled person according to the plan. This activity is outsourced. The plan is agreed upon by both parties. The plan includes the treatment, frequency, and bait plan. (Refer bait plan).

4.7.6 Treatment records, trend reports, plant layouts, bait pans, corrective measures are taken shall be kept as documented information.

5.0 References

5.1 Pest treatment plan
5.2 Bait plan

6.0 Records

6.1 Checklist for Pest control programme – FSMS 04/ PRP/07/CL/01
6.2 Contractor's verification record

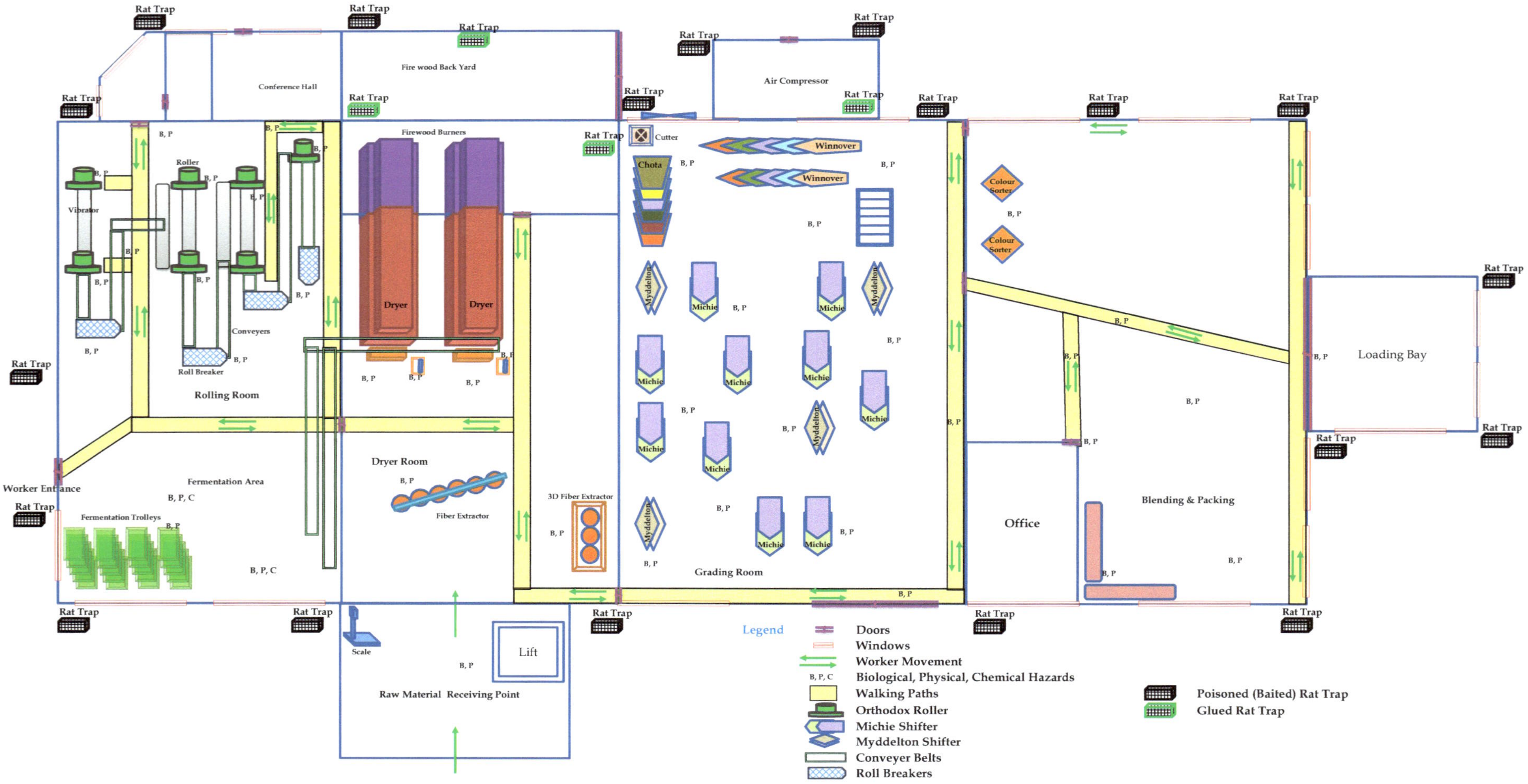
Rat Trap
Conference Hall
Fire wood Back Yard
Firewood Burners
Air Compressor
Cutter
Winnover
Chota
Vibrator
Roller
Conveyers
Roll Breaker
Rolling Room
Dryer
Dryer Room
Fiber Extractor
3D Fiber Extractor
Myddelton
Michie
Grading Room
Colour Sorter
Office
Blending & Packing
Loading Bay
Worker Entrance
Fermentation Area
Fermentation Trolleys
Scale
Lift
Raw Material Receiving Point
B, P
B, P, C
Legend
Doors
Windows
Worker Movement
Biological, Physical, Chemical Hazards
Walking Paths
Orthodox Roller
Michie Shifter
Myddelton Shifter
Conveyer Belts
Roll Breakers
Poisoned (Baited) Rat Trap
Glued Rat Trap

PEST CONTROL CHECKLIST

Activities:
Establish a pest control logbook; maintain the record of inspections and treatment. (It may be necessary to employ a pest control operator to monitor the premises.)

To assist in the prevention of pests, ensure the premises are properly constructed and maintained. For example, seal all unnecessary gaps around fittings or in walls or floors and provide adequate space around equipment to enable easy cleaning.

Food shall be stored so that pests do not have access to it.
Spills shall be removed immediately, and cleaning must be effective.

Vegetation around buildings, including in and around drains, shall be kept under control to discourage pests.

Domestic animals shall not be permitted on the premises.

If pests enter the premises, the action shall be taken to remove them. These measures include the use of baits. Baits are potentially toxic, and care shall be taken during positioning, storage, or handling to avoid contamination of food.

Suitable protective clothing shall be worn when handling pest control chemicals.

Pest of Concern	Products Used	Areas to be checked	Frequency	Responsibility
Mice	Traps, Baits, Gum Boards	Traps, rear yard, seal all gaps	Fortnightly	Q A Officer/FSTL
Cockroach	Fumigation	Throughout	Fortnightly	Pest Controller
Ants	Ant Rid	All surfaces	Daily	Q A Officer/FSTL
Files	Insectocuter	Throughout	Daily	Q A Officer/FSTL
Birds	Bird Traps	Roof	Fortnightly	Q A Officer/FSTL

PEST CONTROL CHECKLIST

Pest Control Checklist: To be completed by Pest Control company representative and Factory Manager or Administrative Officer together after an accompanying visit of the factory premises and related infrastructure.

Name of Pest Controller:					Contact Person:			
Address:						Phone:		
Date	Areas Checked	Follow up Remarks & Future Activities	Evidence of Infestation	Baits Laid Yes/No	Chemicals Applied	Frequency	Responsibility	Verification

ISO 22000 - PREREQUISITE PROGRAMME

PRP 08 – SANITARY AND EMPLOYEE FACILITIES

Distribution:

1. Director/General Manager
2. Factory Manager
3. Factory Officer
4. Quality Officer
5. Assistant Factory Officers
6. Consultant – FSMS

1.0 Objective

This PRP is to ensure adequate sanitary and employee facilities are in place in the factory.

2.0 Scope

This procedure covers sanitary facilities provided to employees and the general sanitation of the factory.

3.0 Responsibility

3.1	General Manager	– Guidance and compliance with factory requirements
3.2	Factory Manager	– Installation and maintenance of equipment
		– Providing required uniforms and shoes
3.3	Factory Officer	– Deciding layout requirements
		– Identification of appropriate equipment
3.4	Quality Officer	– Identification and maintenance of facilities
3.5	Consultant – FSMS	– Ensuring the FSMS requirements are fulfilled

4.0 Activities

4.1 Sanitation facilities and changing facilities shall be available to all employees and staff before they enter production areas. The changing facilities shall accommodate their personal shoes and personal clothes. Space to change clothes and to hang outdoor clothes shall be provided. These shall be so designed that they cannot be used as a general storage area. The changing/locker area shall be large enough to ensure that conditions are not cramped when it is being used at the changeover of shifts. These changing facilities shall be cleaned on a daily basis and shall be inspected by an assistant factory officer at least once a shift.

4.2 Separate toilet facilities shall be provided for male and female employees in the break area. Each restroom shall be equipped with double doors opening inward and well-ventilated. Washroom facilities, toilets, and bathrooms should be hygienically designed. E.g., floors and walls shall be impervious to water, have coved junctions between walls and floor, properly ventilated. The ventilation pipes shall be station away from the food production areas. A plentiful supply of potable water, soap, dispensers of hand sanitizing solution, and clean paper tissue shall be provided in toilets. An adequate number of toilets must be provided. The ideal ratio of toilets to employees shall be 1:25 and washbasins 1:20

4.3 The toilets and bathrooms shall be situated facing away from the production floor, appropriately facing in the direction of the wind. The effluents from the toilets and bathrooms shall be contained in a sealed drainage system and shall not expose to air.

4.4 The toilets shall be washed and disinfected twice a day or more often as required. These toilets shall be inspected at least once a day by a responsible staff for cleanliness and hygiene. (Cleaning Schedule for changing rooms, toilets, and bathrooms, Cleaning checklist for changing rooms, toilets, and bathrooms)

4.5 Eating areas and smoking areas shall be provided since these activities are prohibited in food or packaging processing areas. These areas shall be well maintained and litter-free. The employee canteen and tea rooms (eating areas) smoking areas shall be cleaned after each meal, and they shall ensure no solid waste or any garbage from the canteen is kept open.

4.6 All employees shall be provided with working uniforms and laundry facilities. In addition to the uniform, they shall be provided with aprons when they come in to contact with the product. All uniforms and protective garments shall be changed at least daily or more frequently if required.

4.7 Disposable caps, masks, and gloves shall be provided where necessary. If used, reusable gloves shall be washed every day as a worker leaves the work, and a washed and disinfected pair shall be wearied upon re-entering.

4.8 Hand washing; shall be applied as both operative and management requirement. Accessible and easily operated hand washing and drying facilities shall be provided at the entrance of production areas. Suitable detergents and sterilizing agents shall be made available in dispensing units with hot and cold running water, disposable towels, and properly managed waste receptacles. Water taps shall be knee, elbow sensor operated.

4.9 The handwashing stations shall be inspected by a quality control representative daily for adequate supplies before operation begins. Employees shall be instructed to inform their supervisor when any of the supplies need replenishing. These activities shall be recorded on a daily sanitation audit form.

4.10 First aid kits shall be available in each processing area together with people who have received first aid training. The kits shall be clearly displayed and labelled.

4.11 Litter receptacles shall be placed throughout the factory for litter and garbage, which are adequate for their anticipated loading. These receptacles shall be emptied according to the cleaning schedule. Personnel engaged in the emptying of litter receptacles in processing areas shall not be employed outside.

5.0 References

5.1 Cleaning Schedule for changing rooms, toilets, and bathrooms.

6.0 Records

6.1 Cleaning checklist for changing rooms, toilets, and bathrooms – FSMS 04/PRP/08/CL/01

6.2 Restroom cleaning record – FSMS 04/PRP/08/RC/01

Suitability Check List for Changing Rooms/Toilets/Bath Rooms							
Plant Name				Department			
D. Ref	FSMS 04/PRP/08/CL/01		Responsible Officer				
Employee Facility	Damages to Amenities	Location	Date Inspected	Hygienic Condition	Cleaning Quality	Suitability	Remarks

FSMS 04/PRP/08/RC/01

Rest Room Cleaning Record

Month		Year	

Date	All units are free of debris & needless utensils	Dining hall is properly cleaned	Garage & outside of restroom is properly cleaned	Female restroom is cleaned daily	Female Restroom - Floor, doors, windows, ceiling are properly cleaned & free of cobwebs or dust	Male restroom is cleaned daily	Male Restroom - Floor, doors, windows, ceiling are properly cleaned & free of cobwebs or dust	Male washroom cleaned daily	Female washroom is cleaned daily	Kitchen & fireplace cleaned daily	All light bulbs/equipment are cleaned and operational	All lockers are cleaned daily	Waste bins are cleaned daily	**Reasons for Not Performing**	**Carry out Later**	**Withering Operator for the Day**	**Checked By(A F O - Withering)**	**FSTL's Comments**
01																		
02																		
03																		
04																		
05																		
06																		
07																		
08																		
09																		
10																		
11																		
12																		
13																		
14																		
15																		
16																		
17																		
18																		
19																		
20																		
21																		
22																		
23																		
24																		
25																		
26																		
27																		
28																		
29																		
30																		
31																		

Tick the Relevant Cage with ☑ or ☒

ISO 22000 - PREREQUISITE PROGRAMME

PRP 09 – SUPPLY OF AIR, WATER, AND ENERGY

Distribution:

1. Director/General Manager
2. Factory Manager
3. Factory Officer
4. Quality Officer
5. Assistant Factory Officers
6. Consultant – FSMS

1.0 Objective

The objective of this procedure is to ensure an uninterrupted, hazard-free supply of compressed air, water, and energy.

2.0 Scope

The scope covers the supply of compressed air, water, and energy for the production process and cleaning.

3.0 Responsibility

3.1 General Manager – Identification of necessary equipment and utensils

3.2 Factory Manager – Installation of necessary equipment and utensils

3.3 Factory Officer – Uninterrupted supply of utilities

3.4 Quality Officer – Facilitation

3.5 Asst. Factory Officers – Facilitation

3.6 Consultant – FSMS – Confirmation of suitability

4.0 Activities

4.1 Air

4.1.1 Air used as raw material, or direct contact with food shall be odourless, free of compressor lubricants and water, or moisture.

4.1.2 Air shall be passed through filters to prevent contamination.

4.2 Water

4.2.1 The water used as a raw material for food processing shall be confirmed to WHO standards or SLS: 614 (Potable water). Water shall be tested regularly, and records shall be maintained from the accredited laboratory regularly. (Testing schedule for water)

4.2.2 The water used for other operations such as cleaning of equipment and other utensils (which act as a food contact surface) shall be confirmed to SLS: 614 (Potable water).

4.2.3 Work instruction for the handling of the Water Treatment unit.

4.2.4 Water lines carrying potable water and raw water shall be marked with colour codes for easy identification.

4.2.5 Water used for personal hygiene activities, general cleaning activities shall be UV filtered or chlorinated, and a residual chlorine level of 0.02-0.1ppm shall be maintained.

4.2.6 Additional water storages shall be maintained to prevent impact from essential service breakdowns.

4.3 Energy

4.3.1 An uninterrupted flow of energy shall be assured through the installation of power generators to prevent and minimise the impact on food safety.

4.3.2 An adequate supply of firewood and proper storage facilities shall be provided to prevent and minimise the impact on food safety.

5.0 References

5.1 SLS: 614 (Potable water).

5.2 Work instruction for Water Treatment

6.0 Records

6.1 Water inspection and test reports – FSMS 04/PRP/09/RP/01

6.2 Material safety data sheets - Chemicals used

ISO 22000 - PREREQUISITE PROGRAMME

PRP 10 – WORK ENVIRONMENT

Distribution:

1. Director/General Manager
2. Factory Manager
3. Factory Officer
4. Quality Officer
5. Assistant Factory Officers
6. Consultant – FSMS

1.0 Objective

This PRP outlines the establishment, management, and maintenance of providing a clean and safe work environment.

2.0 Scope

This procedure covers the processing and storage areas of orthodox black tea manufacture. This also covers the handling of refuse tea and appropriate disposal according to SLTB regulations.

3.0 Responsibility

3.1	General Manager	– Providing Guidance and establishing requirements
3.2	Factory Manager	– Providing necessary facilities
		– Maintenance of the facilities
3.3	Quality Officer	– Identification of necessary requirements
3.4	Asst. Factory Officers	– Monitoring the effectiveness
3.5	Consultant – FSMS	– Ensuring the FSMS requirements are fulfilled

4.0 Activities

4.1 Processing and storage areas shall be separated, and cross movements shall be restricted.

4.2 Pathways shall be demarcated to prevent cross-contamination.

4.3 A sufficient supply of fresh air shall be provided to each processing area, depending on the operation.

4.4 Necessary resources shall be disposed of as and when required according to the situation to provide a clean and hassle-free work environment for the employees.

4.5 The management shall ensure to create non-discriminatory, non-confrontational working conditions to reduce environmental stress to obtain effective work attitude from the employees.

4.6 The humidity, temperature, heat light, airflow, and noise shall be controlled, or protection measures will be provided when required.

4.7 The employees shall be fairly treated according to the human resource policies implemented.

4.8 Relevant occupational health and safety measures shall be implemented, including fire safety, smoke protection, maximum weight limits including protective uniforms, goggles and gumboots, and first aid whenever applicable.

4.9 The sickroom and relevant infrastructure shall be provided with on-call facilities for medical treatments in case of an emergency.

4.10 Floors of processing areas shall be maintained clean & dry during production time.

4.11 Walls, roofs, ceilings & overhead fixtures shall be cleaned regularly to prevent the accumulation of dirt and dust.

4.12 All solid waste shall be collected in closed bins and removed from the factory to the enclosed collection point regularly to prevent accumulation.

4.13 The collected waste shall be disposed of with care to avoid cross-contamination and multiplication.

4.14 All internal drains shall be maintained clean to prevent stagnation of water.

4.15 All external drains shall be cleaned once a month to prevent water stagnation.

4.16 External housekeeping activities shall be supervised by the Administrative Officer.

4.17 Environmental monitoring (microbiological tests) shall be conducted according to the testing schedule to evaluate the level of air-bone contamination.

4.18 Necessary improvements and upgrades shall be made based on the findings of periodic evaluations.

4.19 Continuous communication between officials of the core production processes and support services shall be ensured.

4.20 The refuse tea released to the environment shall be confirmed to the legal requirements.

4.21 The functionality and effectiveness of waste disposal shall be monitored by the Factory Manager.

4.22 Solid waste shall be disposed of according to the statutory and regulatory requirements.

4.23 Sewage disposal shall comply with sanitary and regulatory requirements specified by the regional health inspection services.

4.24 Facilities and methods of disposal of the contracted refuse tea collectors (if any) shall be monitored and approved by the company.

4.25 The administration buildings and peripheral areas shall be kept organized and well cleaned, where separate cleaning staff shall be hired as necessary.

5.0 References

5.1 Testing Schedule for environmental monitoring.

5.2 Refuse tea disposal methods and schedules of SLTB

6.0 Records

6.1	Test report – environmental monitoring	– FSMS 04/PRP/10/RP/01
6.2	Refuse tea disposal record	– FSMS 04/PRP/10/RC/01
6.3	Cleaning checklist for admin and peripheral areas	– FSMS 04/PRP/10/RC/02

D. Ref: FSMS 04/PRP/10/RC/01

Administration building and Peripheral Area Cleaning Record

Month		Year	

Date	Outside areas are free of debris & needless utensils	Main Road & adjoining areas are properly cleaned	Main Gate & adjoining area is properly cleaned	Security Hut is properly cleaned	Security Hut - Floor, doors, windows, ceiling are properly cleaned & free of cobwebs or dust	Office is properly cleaned	Office - Floor, doors, windows, ceiling are properly cleaned & free of cobwebs or dust	Factory front is properly cleaned	Backyard & sides of the Factory is properly cleaned	Firewood store & kiln are properly cleaned	All light bulbs/equipment are cleaned and operational		Conference Hall is properly cleaned	Waste bins are cleaned daily	**Reasons for Not Performing**	**Carry out Later**	**Withering Operator for the Day**	**Checked By(A F O - Withering)**	**FSTL's Comments**
01																			
02																			
03																			
04																			
05																			
06																			
07																			
08																			
09																			
10																			
11																			
12																			
13																			
14																			
15																			
16																			
17																			
18																			
19																			
20																			
21																			
22																			
23																			
24																			
25																			
26																			
27																			
28																			
29																			
30																			
31																			

Tick the Relevant Cage with ☑ or ☒

ISO 22000 - PREREQUISITE PROGRAMME

PRP 11 – CLEANING AND SANITATION OF PLANT, MACHINERY, EQUIPMENT, AND UTENSILS

Distribution:

1. Director/General Manager
2. Factory Manager
3. Factory Officer
4. Quality Officer
5. Assistant Factory Officers
6. Consultant – FSMS

1.0 Objective

To prevent contamination of food and beverages manufactured due to un-cleaned plant, machinery equipment, and utensils.

2.0 Scope

The scope covers cleaning and sanitation of the plant, machinery, equipment, and utensils, that come into direct contact with food. This procedure also covered the cleaning before the commencement of production, mid-shift cleaning, and cleaning after production.

3.0 Responsibility

3.1 General Manager – Providing Guidance and establishing requirements

3.2 Factory Manager – Providing necessary facilities

3.3 Factory Officer – Guidance, facilitation, and execution

3.4 Quality Officer – Facilitation, execution & verification

3.4 Asst. Factory Officers – Execution

3.5 Consultant – FSMS – Ensuring the FSMS requirements are fulfilled

4.0 Activities

4.1 The plant, machinery, equipment, and utensils that come into direct contact with food being processed shall be cleaned or sanitized as required prior to commencement of production.

4.2 The plant, machinery, equipment, and utensils shall be cleaned immediately before the commencement of production.

4.3 The plant, machinery, equipment, and utensils shall be cleaned immediately after production (manual cleaning) to ensure no residues of food materials or cleaning agent residues remained.

4.4 All the in-process cleaning shall be carried out on time by responsible personnel to ensure proper food safety of the products manufactured.

4.5 Buildings, fixtures, and other physical facilities shall be repaired and maintained in stipulated sanitary conditions as to the work instructions provided to minimize products from becoming unintendedly adulterated.

4.6 Cleaning and sanitizing of machinery, equipment, and utensils shall be conducted to protect against contamination of made tea, intermediate product, raw materials, product contact surfaces, or packaging materials.

4.7 All the contact surfaces, including product contact surfaces of equipment and utensils, shall be cleaned as frequently as recommended to protect against contamination.

4.8 The contact surfaces of the firing room, grading room, blending, and packing area shall be kept under low moisture levels, while all the contact surfaces shall be in a dry, sanitary condition at the time of use.

4.9 If the surfaces are required to be wet cleaned while production on the move in areas mentioned in 4.8, check the possibility of using IPL, or if surfaces are wet-cleaned, they shall be sanitized and thoroughly dried prior to subsequent use.

4.10 The rolling room operation shall be protected against the introduction of microorganisms into rolled tea leaves. Hence all the contact surfaces shall be cleaned and sanitized prior to use and after any interruption during which the product contact surfaces may have become contaminated.

4.11 Machinery, equipment, and utensils that are used in a continuous production shall be cleaned and sanitized as a clean as you go operation.

4.12 If any single service article is intended for use, it shall be stored in appropriate containers and handled, dispensed, used, and disposed of in an appropriate manner to protect against contamination of product and product contact surfaces.

4.13 The equipment shall be cleaned and sanitized on a regular schedule as planned to ensure the wholesomeness of the product by reducing bacterial loads, chemical or physical contaminants, where cleaning procedures shall be followed as outlined in the relevant work instructions. The equipment, utensils shall be stored or rested in a sanitary manner when they are not in use.

4.14 Unessential items that are not necessary for day-to-day operations shall be kept in the right places following the 5S concepts/instructions to reduce poor housekeeping practices as well as to improve proper food hygiene inside the facility. The cleaning utensils, including brooms, squeegees, and mops, shall be kept in specific hangars in their storage places off the floor.

4.15 Any idling equipment, small articles such as pencils, knives, fused, tools, etc., shall be removed from processing areas as well as they shall not be left in or on electrical panels, switchboards, processing equipment, etc., and stored in appropriate places in a clean, hygienic manner.

4.16 The obsolete equipment, utensils shall not be allowed to accumulate inside the production facility, which should be immediately removed from the production and store in the right places or dispose of as appropriate.

4.17 Cleaned and sanitized equipment and utensils that are moveable shall be stored in a manner that protects contact surfaces from contamination.

4.18 Steel wool or metal sponges shall not be used in the cleaning of machinery, equipment, or utensils.

4.19 Any reusable ingredient container or utensil shall be thoroughly cleaned and dry before re-use.

4.20 Cleaning compounds and sanitizing agents used in cleaning and sanitizing procedures shall be free from undesirable microorganisms and be of recommended food grade that is absent of odour or colours, as well as adequate and safe for conditions of use.

4.21 Thermometers, recording charts, and pressure gauges, or any other measuring devices shall be provided where applicable, which must be in good working order, calibrated, and easily accessible. Where applicable, each piece of equipment shall have written procedures for complete dissection and cleaning.

4.22 All cleaning activities shall be monitored and recorded by personnel responsible for monitoring as mentioned in the cleaning schedule.

4.23 Cleaning swabs shall be from random equipment at random places from time to time and send to the laboratories for the verification of cleaning and sanitizing activities, which shall be carried out by the QA Officer without prior notices.

4.24 The effectiveness of all cleaning and sanitation activities shall be evaluated as per the verification schedule.

5.0 References

5.1 Cleaning schedule.

5.2 Plant cleaning and sanitation work instructions.

Cleaning of the withering area - FSMS 03/WI/WA/01

5.2.1 Cleaning and sanitation of rolling room, rolls, roll breakers, conveyors, and fittings - FSMS 03/WI/RR/02

5.2.2 Cleaning of firing area, conveyers, and fittings - FSMS 03/WI/FR/03

5.2.3 Cleaning of sifting area - FSMS 03/WI/GR/04

5.2.4 Cleaning of sifters, colour sorting machines, winnower, and fittings.

5.2.5 Cleaning of holding and bulk retaining bins.

5.2.6 Cleaning of working tables and trolleys.

5.2.7 Cleaning of Crates.

5.2.8 Cleaning of measuring equipment.

6.0 Records

6.1 Monitoring and verification records in each specific area

Withering Area	- FSMS 04/PRP/11/RP/01
Rolling Room	- FSMS 04/PRP/11/RP/02
Dryer Room	- FSMS 04/PRP/11/RP/03
Sifting Room	- FSMS 04/PRP/11/RP/04

6.2 Cleaning checklist on machinery, equipment and utensil

- FSMS 04/PRP/11/CL/01

Cleaning Check List for Machinery, Equipment, and Utensil							
Plant Name					Department		
D. Ref	FSMS 04/PRP/11/CL/01		Responsible Officer				
S. No	Machine/Equipment/ Utensil	Point of Use	Date Inspected	Hygienic Condition	Cleaning Quality	Suitability	Remarks

Withering Area Cleaning Record

Month		Year		Loft No	

Date	All units are free of debris & needless utensils	Staircase is properly cleaned	Loft floor is properly cleaned	All trough fans are adequately cleaned	All withering troughs & nets are adequately cleaned	Top & underneath of all withering troughs are adequately cleaned	All doors, windows & ceiling are appropriately cleaned & free of cobwebs or dust	All Window nets are properly washed & cleaned	All light bulbs/equipment are cleaned and operational		Leaf release hoppers are adequately cleaned	All hygrometers are cleaned & operational	Waste bins are cleaned daily	Reasons for Not Performing	Carry out Later	Withering Operator for the Day	Checked By(A F O – Withering)	FSTL's Comments
01																		
02																		
03																		
04																		
05																		
06																		
07																		
08																		
09																		
10																		
11																		
12																		
13																		
14																		
15																		
16																		
17																		
18																		
19																		
20																		
21																		
22																		
23																		
24																		
25																		
26																		
27																		
28																		
29																		
30																		
31																		

Tick the Relevant Cage with ☑ or ☒

Prepared	: FSTL	Revision	: 00
Reviewed	: C - FSMS	Date	: 01-11-2018
Approved	: D/GM	Page	: Page 1 of 1

Rolling Room Cleaning Record

Month		Year	

Date	All units are free of debris & needless utensils	All the parts of conveyors & vibrators are adequately cleaned	Humidifiers are properly cleaned	Floor & the walls of the rolling room is properly cleaned	Staircase is properly cleaned	Trolleys are properly cleaned	Plastic boxes/trays/cleaning equipment are properly cleaned & stored safely	Top & underneath of all the rolls & roll breakers properly cleaned	All doors, windows & ceiling are adequately cleaned & free of cobwebs or dust	All light bulbs/equipment are cleaned and operational	Waste bins are cleaned daily	**Reasons for Not Performing**	**Carry out Later**	**Withering Operator for the Day**	**Checked By(A F O - Withering)**	**FSTL's Comments**
01																
02																
03																
04																
05																
06																
07																
08																
09																
10																
11																
12																
13																
14																
15																
16																
17																
18																
19																
20																
21																
22																
23																
24																
25																
26																
27																
28																
29																
30																
31																

Tick the Relevant Cage with ☑ or ☒

Firing Area Cleaning Record

Year		Month	

Date	All units are free of debris & needless utensils	Inside & outside of ovens are properly cleaned	Heaters & pipes are adequately cleaned	Drying chamber is properly cleaned	Chimney/chambers are adequately cleaned	Conveyers are properly cleaned	3T machine is adequately cleaned	Fibre mat machine is adequately cleaned	Firewood store is properly cleaned & stacked	Floor is adequately cleaned	Magnets are adequately cleaned	All doors, windows & ceiling are adequately cleaned & free of cobwebs or dust	All light bulbs/equipment are cleaned and operational	Waste bins are cleaned daily	Reasons for Not Performing	Carry out Later	Firing Operator for the Day	Checked By(A F O – Withering)	FSTL's Comments
01																			
02																			
03																			
04																			
05																			
06																			
07																			
08																			
09																			
10																			
11																			
12																			
13																			
14																			
15																			
16																			
17																			
18																			
19																			
20																			
21																			
22																			
23																			
24																			
25																			
26																			
27																			
28																			
29																			
30																			
31																			

Tick the Relevant Cage with ☑ or ☒

Grading Area Cleaning Record

Year Month

Date	All units are free of debris & needless utensils	Middletons are adequately cleaned				Michi's are adequately cleaned										Chota is adequately cleaned	Winnowers are adequately cleaned		Colour separators are adequately cleaned			Cutter machine is adequately cleaned	Floor is adequately cleaned	All doors, windows & ceiling are properly cleaned & free of cobwebs or dust	All light bulbs/equipment are cleaned and operational		Waste bins are cleaned daily	Reasons for Not Performing	Carry out Later	Grading Operator for the Day	Checked By(A F O - Withering)	FSTL's Comments
		1	2	3	4	1	2	3	4	5	6	7	8	9	10		1	2	1	2	3				1	2						
01																																
02																																
03																																
04																																
05																																
06																																
07																																
08																																
09																																
10																																
11																																
12																																
13																																
14																																
15																																

Black Tea

XYZ TEA FACTORY
ISO 22000 - PREREQUISITE PROGRAMME

D. Ref: FSMS 04/PRP/11/RP/04

Date	All units are free of debris & needless utensils	Middletons are adequately cleaned				Michis are adequately cleaned										Chota is adequately cleaned	Winnowers are adequately cleaned		Colour separators are adequately cleaned			Cutter machine is adequately cleaned	Floor is p adequately cleaned	All doors, windows & ceiling are adequately cleaned & free of cobwebs or dust	All light bulbs/equipment are cleaned and operational		Waste bins are cleaned daily	Reasons for Not Performing	Carry out Later	Grading Operator for the Day	Checked By(A F O - Withering)	FSTL's Comments
		1	2	3	4	1	2	3	4	5	6	7	8	9	10		1	2	1	2	3				1	2						
16																																
17																																
18																																
19																																
20																																
21																																
22																																
23																																
24																																
25																																
26																																
27																																
28																																
29																																
30																																
31																																

Tick the Relevant Cage with ☑ or ☒

ISO 22000 - PREREQUISITE PROGRAMME

PRP 12 – SUPPLIER EVALUATION AND SELECTION

Distribution:

1. Director/General Manager
2. Purchasing Manager
3. Factory Manager
4. Factory Officer
5. Quality Officer
6. Stores Manager
7. Stores Executive
8. Consultant – FSMS

1.0 Objective

To provide a transparent methodology with uniform consistency for the purchase of products or services, which shall ensure the fulfilment of relevant food safety/quality assurance requirements while establishing appropriate control measures to eliminate food safety hazards through incoming materials or services.

2.0 Scope

This procedure applies to all material purchases and services, including sourcing, supplier selection, and evaluations, maintaining purchasing up-to-date, development of alternative sources of purchasing while establishing, maintaining, and upgrading the purchasing system.

3.0 Responsibilities

3.1	General Manager	– Guidance and compliance with requirements
3.2	Purchasing Manager	– Selection & evaluation of new suppliers – Periodical review of existing suppliers – Inventory control, procurement planning & follow-ups
3.3	Factory Manager	– Identification of materials & Service requirements
3.4	Consultant	– Compliance with FSMS
3.5	Stores Manager	– Maintaining & updating inventory controls – Receiving, inspection & documentation of correct items – Storage & issue of materials according to FIFO principles – Supervision of inventory and record-keeping
3.6	Stores Executive	– Collection of samples, supervision of material storing – Record keeping and issuing of materials

4.0 Activities

4.1 Supplier Evaluation & Selection

4.1.1 The material and service requirements shall be planned according to the production planning. The forecasted raw material requirements shall be categorized and listed while calling suppliers based on the company registered list of suppliers or by tendering.

4.1.2 The primary raw material suppliers are tea small-holder farmers and collectors.

4.1.2.1 Thus, selection, evaluation, registration is carried out by the purchasing team and shall be trained by the food safety team. Individual suppliers and collectors (through their network of individual suppliers) shall maintain a field record book for each specific field that contains the information of plant phenotype (clone), year of planting, pruning dates, fertilizer application dates, weeding dates, or weedicide application dates and the fluking records including an individual employee with quantities harvested, field codes, etc.

4.1.2.2 If farmers are unable to provide required information on an unannounced visit to any specific field, the supplier shall be discontinued temporarily until the corrective actions are properly implemented and verified, and the farmer or farmer groups in lack of compliance shall be retrained and frequently visited until such conditions are corrected and evaluated.

4.1.3 The suppliers shall provide all the relevant details, including but not limited to; material safety data sheets, product descriptions, specific certifications (ISO 22000, ISO 9001, GMP, etc.), samples, and other relevant data which may be specified by the production team time to time.

4.1.4 The factory visits or external audits shall be performed as to the criticality of the materials where supplier shall be responsible for third party audits if required while letting the purchasing team to make observations on their production facilities as well as up to the primary raw material receiving of any specific product manufactured by a given supplier.

4.1.5 Suppliers shall be selected based on previous ratings, capacity, and quality of the materials supplied, and the cost.

4.1.6 Each item shall be furnished with the best suppliers, or when material scarcity or cost is too high, alternative suppliers and temporary suppliers shall be contacted according to the list of suppliers.

4.2 Purchasing of Materials

4.2.1 After selection and evaluation, the new suppliers shall be given with trial orders, and their performances are evaluated; the second and further continuation of their service shall be based on the performance of the given orders.

4.2.2 Each such supplier is evaluated after some time till sufficient information shall be collected before they become regular suppliers.

4.2.3 Each supplier shall be rated according to their performance, and the status of the supplier (priority, secondary, alternative, or temporary) shall be offered according to the ratings.

4.3 Approved Supplier List

4.3.1 The approved supplier list shall be maintained and updated continuously for each item and service required.

4.3.2 Each item shall be listed with relevant specific requirements to be fulfilled by the supplier and the relevant chemical, microbiological and physical hazards associated with and their maximum levels and testing reports to be attached.

4.3.3 Each such item supplies are closely monitored, and relevant suppliers are educated according to the requirements and the understanding of the suppliers.

4.3.4 The checklists shall be maintained with verification for specific criteria for each item with food safety hazards, supplier control measures, product specifications, and the communication on such parameters with suppliers.

4.3.5 The key suppliers shall be selected based on their capacity to comply with specific requirements, on-time delivery, material quality, the effectiveness of service delivery, and especially food safety.

4.4 Supplier Rating

4.4.1 The supplier evaluation and review shall comply with; in-terms of their production facilities, customer lists, the capability for on-time delivery of specific products to comply with procurement parameters.

4.4.2 Trail order performance on given supplies and the evaluation of the product quality.

4.4.3 Market leadership, reputation, and dealerships for a safe and sound food product or raw materials supply over a sufficient period of time.

4.4.4 Any other parameters such as monopolistic suppliers who dominate the markets.

4.4.5 All the suppliers and contractors who filtered through the above selection criteria shall be given appropriate ratings and shall be included in the approved supplier list with adequate priority based on the ratings.

4.4.6 If a new supplier has to be appointed, he shall come through the same procedure passing all the set parameters to be listed in the approved supplier list.

4.5 Purchasing Data

4.5.1 Raw material specification data sheets shall be prepared according to the ISO 22000 requirements. The same specifications shall be utilized for the purchasing requirements and provided to suppliers as a form of purchase order requirements.

4.5.2 Following requirements shall incorporate into the purchase orders according to the type of product or materials purchased:

- The product name, type, brand name, or any other specific product identification
- Material specification sheets, packaging, quantity, mode of dispatch
- The specifications documented for all raw materials and ingredients
- Requirements of the food-grade materials, source of raw materials, and food safety requirements specified.
- Regulatory requirements where applicable
- Delivery requirements and costs where necessary
- Product specifications for test reports, relevant certifications, method of product approvals

4.5.3 The purchasing data shall document as described to ensure that all purchase data such as quantity, quality, food safety requirements, and delivery methods with dates. It is clearly documented and reviewed before approval of purchase order. If any guarantee certificate is required, the same requirement shall be communicated to the supplier.

4.6 Purchasing of Services

4.6.1 Purchase of services such as calibration of equipment, instruments, maintenance equipment, pest control, fumigation, testing of products, shelf life, transportation, including air freight and marine, insurance, etc., shall be carried out through approved and accredited service providers.

4.6.2 In such events, instead of issuing purchase orders, relevant heads of departments directly deal with the relevant approved service provider, while purchasing manager shall maintain selection and evaluation with the help of relevant department heads for selection and evaluation of their services.

4.7 Control of Suppliers

4.7.1 The type of the extent of control shall depend on the type of product or raw material purchased. The supplier's performance of the previous deliveries, impact of the products/materials being procured on the quality of the final product, the quality, and food safety compliance records supplied with the delivery.

4.7.2 The control measures shall be specified in the purchase order or in the contract.

4.7.3 All the raw materials/products/ingredients shall be tested at the factory after delivery according to the illustration given. If any material is rejected due to the failure of compliance with given specifications, the product shall be delivered to the supplier through his own cost of transportation.

4.7.4 The external or internal laboratory test shall be conducted by the company as the main control measure for all chemicals, raw materials, ingredients, etc.

4.7.5 All the necessary records related to purchasing and individual suppliers as well as service providers shall be prepared and maintained up to date.

5.0 References

5.1 Raw material Specifications

5.2 PRP 02 – Management of Purchased Materials

5.3 Process Flow Diagram for Supplier Evaluation – FSMS 04/PRP/12/PFD/01

6.0 Records

6.1 Supplier Registration Form

6.2 Indent and Incoming Material Inspection Report

6.3 Approved supplier/Contractor List

1. Raw Materials & Ingredients
2. Services

6.4 Supplier Performance Records

6.5 Supplier Ratings

6.6 Product Hygiene Records and Supplier Performance Check List

Process Flow Diagram – Supplier Evaluation – FSMS 04/PRP/02/PFD/01

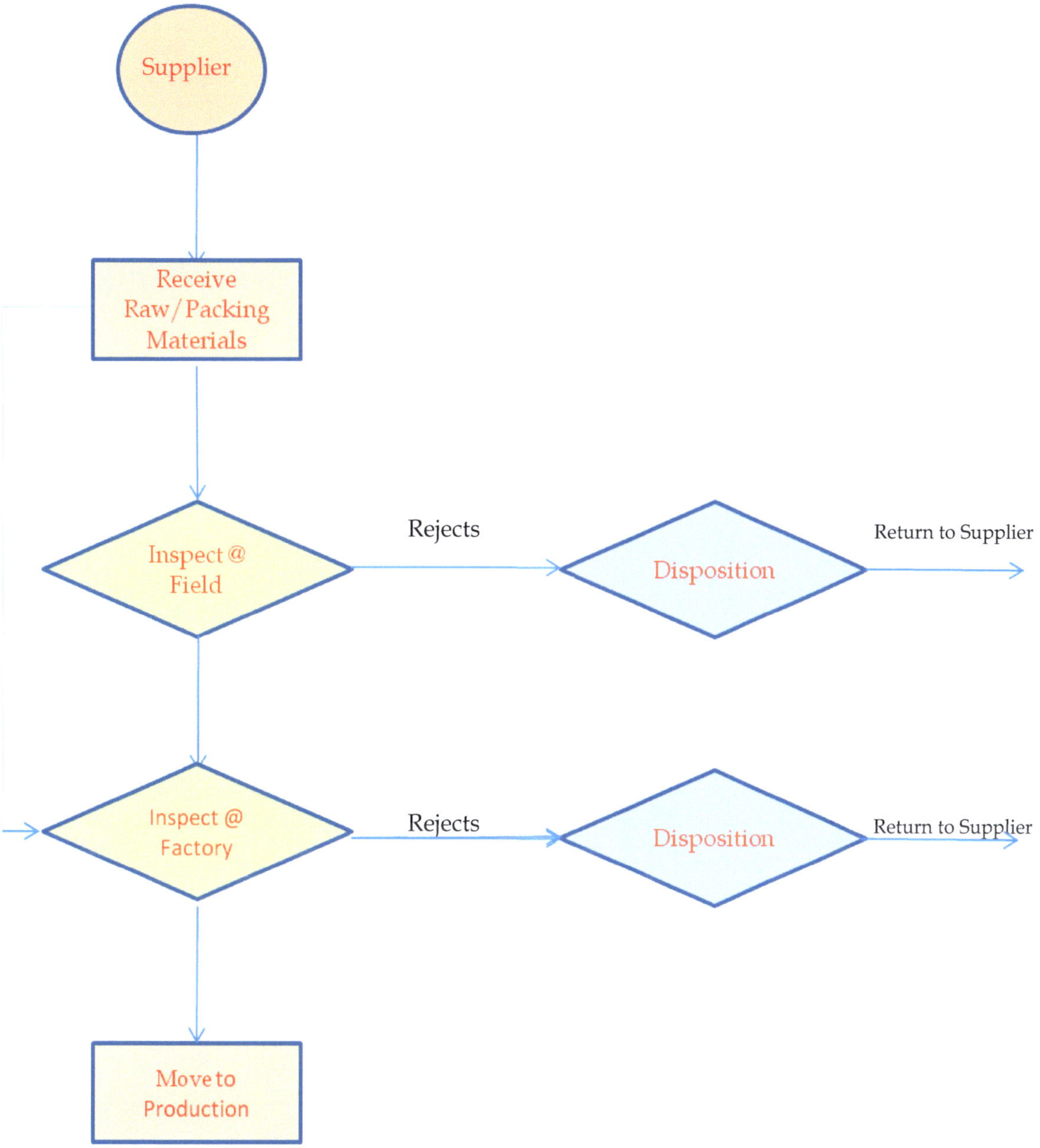

Black Tea

XYZ TEA FACTORY
SUPPLIER EVALUATION & SELECTION

D. Ref: FSMS 04/PRP/12/RC/01

LEAF COUNT – GREEN LEAF QUALITY INSPECTION RECORD

No	Time of Receiving	D. K. No.	Supplier		Leaf Quality			Water %	*Amount Over 22Kg	Damaged Leaf (Kg)	Date		FSTL's Comments (Verification)
			Name	Rank	B	BB	P				Remarks	Signature	
01		FACTORY	XYZ T/F										
02		FACTORY	LINE 01										
03		FACTORY	LINE 02										
04		FACTORY	LINE 03										
05		FACTORY	LINE 04										
06		GL 4526	MR. N. V. AJANTHA										
07		GL 5309	MR. M. V. P. A. SHANTHA										
08		GL 4341	MR. K. A. JINADASA										
09		GL 5460	MR. D. L. THARANGA										
10		GL 2956	MR. M. P. SUNIL										
11		GL 4462	MR. K. W. NANDADASA										
12		GL 5497	MR. E. K. J. BANDARA										
13		DM 3987	MR. M. G. NISHANTHA										
14		GL 4874	MR. W. R. N. P. KUMARA										
15		M 4393	MR. T. P. G. DAMITHA										
16		T/S/H/D/S	ULUWITIKE										
17		T/S/H/D/S	MEEPAWALA										
18			MR. SUGATHAPALA										
19			MR. NALIN										
20			MR. JANAKA										
21													
22													
PERCENTAGE %													

*package gross weightage over 22Kg of Gunny Sacks or Creates

....
Signature
Quality Officer

....
Signature
Factory Manager

....
Signature
Managing Director

Prepared : FSTL
Reviewed : C - FSMS
Approved : D/GM

Revision : 00
Date : 01-11-2018
Page : Page 1 of 1

ISO 22000 - PREREQUISITE PROGRAMME

PRP 13 – FOOD DEFENCE

Distribution:

1. Director/General Manager
2. Purchasing Manager
3. Factory Manager
4. Factory Officer
5. Quality Officer
6. Stores Manager
7. Administrative Officer
8. Security Officer
9. Stores Executive
10. Consultant – FSMS

1.0 Objective

To quickly and accurately identify, respond to, and contain threats or acts of intentional contamination strictly controlled through planned arrangements, requirements with the existing food safety management system, and is being effectively implemented and updated continuously.

2.0 Scope

This procedure covers all the security measures applied including, risk or threat assessments, annual evaluations, employee screening, access control, and inspection related to food defence and food safety.

3.0 Responsibilities

General Manager	- Overall
Factory Manager	- Monitoring, maintain data collection, guidance, team management (TL)
Security Officer/Consultant	- Security requirements
Quality Assurance Manager	- Maintain food defense system
Food defense team/FST	- Maintaining/updating/documenting preparation of relevant documents self-assessment, audit, corrective actions

4.0 Activities

4.1 General

4.1.1 The XYZ Pvt. Ltd., who manufacture, process, and pack orthodox black tea, shall have the following food defence procedure to deter and prevent intentional contamination and will have protocols in place to quickly and accurately identify, respond to, and contain threats or acts of intentional contamination. Nonetheless, a company registered green leaf collectors and individual suppliers will ensure their suppliers adopted to given protocols and implement appropriate controls, which will be audited annually.

4.1.2 The laws and legislation requirements governing food defence vary from country to country. Hence, the company has defined a minimum set of food defense standards to help the company meets legal and consumer expectations, where selected standards may exceed the requirements of a specific country or area.

4.1.3 The management has selected the food safety team as the plant food defence team, where additional consultants on the specific requirement will be hired from the outside.

4.1.4 The food defence plan shall be developed along with the food safety plan covering areas that are not covered during the hazard analysis in line with intentional adulterations on the company's products.

4.1.5 The company's written food defence plan includes vulnerability assessment, including required explanations, to identify significant vulnerabilities and actionable process steps. The process has defined mitigation strategies, including required explanations for food defence monitoring procedures and corrective actions verification plans for the complete assessment of the food defence program.

4.1.6 Statutory and regulatory authorities and other organizations, including clients/distributors/supermarket chains that may have an interest or impact on food defence shall be informed as appropriate on issues concerning them. Such communication shall be done by or under the approval of the General Manager.

4.1.7 Any inquiries, customer feedback, or any other information from tea auction, bulk buyers, or supermarket chains, including customers, suppliers, and statutory and regulatory authorities regarding food defence shall be communicated to Quality Assurance Manager. All above information and issues shall be documented and communicated.

4.1.8 The food defence team has selected mitigation strategies for actionable process steps, which identify and implement mitigation strategies at each actionable process step to significantly minimize vulnerabilities, including a written explanation of how the strategy can minimize the vulnerability.

4.1.9 Food defence monitoring plan shall monitor the mitigation strategies with adequate frequency to provide assurances that they are consistently performed. Verify that mitigation strategies are properly implemented and appropriate decisions about food defence corrective actions are being made and implemented while conducting validation of the food defence plan as appropriate.

4.2 Gap Analysis/Self-assessment

4.2.1 A gap analysis/self-assessment shall be conducted prior to the new food defence program, and every year there will be an annual review of the existing system to identify gaps, new treats, adaptability, suitability, and continuity. Thus, the company can maintain its food defence program by actually fixing gaps identified in assessments and modifying the plans and programs as conditions change. The key activities are;

a. Identify the most vulnerable components of the site's operations and determine how to use resources, procedures, and protocols most effectively to reduce risks;
b. Identify potential areas of exposure or "gaps" in the site's security measures; and,
c. Once the gaps are identified, countermeasures (such as program or process improvements or physical security hardware upgrades) can be used to bridge or fix the gaps.

4.2.2 The system shall be audited under the ongoing internal audit plans based on this procedure. However, the company will hire an external food defence auditor bi-annually with an internal audit team to make sure the system is running at its optimum conditions. The internal audits shall focus on the plans and procedures developed while assessing the site's implementation of the plan, which will show any gaps that need to be fixed.

4.2.3 The food defence system of the company shall be adapted to the PAS 96: 2014 as the main food defence standard that will be used to develop the company's food defense plan and will use the PAS 96's Risk Assessment methods to evaluate its food defence system requirements together with Hazard Analysis of ISO 22000 FSMS.

4.2.4 The company has developed a comprehensive risk assessment checklist for the evaluation of risks together with PSA 96:2014.

4.2.5 Company has established an employee hiring manual, which includes a minimum level of background checks for all the employees.

4.3 Access Control and Visitor Management

4.3.1 Effective access control is the foundation of a successful food defence plan where the company has implemented an automated access control system for the office areas, and production with identification colours for each section, ID cards for employees with their credentials, and limited access visitor passes.

4.3.2 Security shall identify and document all individuals gaining access to the facility, limit access to those who have a legitimate reason to be there, control pedestrian and vehicle traffic at the site, and properly report and respond to incidents challenging this policy.

4.3.3 The visitors and general employees shall be controlled for gaining access to processing and manufacturing, raw material, ingredient, packaging storage areas, hazardous and chemical storage areas, as well as shipping and receiving areas.

4.3.4 Access to those (4.3.3) areas is limited to the individuals who require access in order to complete their assigned tasks, and their uniforms are specifically colour coded relevant to the working area. If a visitor is gaining access for any specific reason, he shall be accompanied by a company executive.

4.3.5 The vehicles coming in and going out of the facility shall be inspected, recorded, and verified before gaining access or leaving, whether they belong to the company, employees, visitors, or suppliers. The parked vehicles shall display a placard or decal issued by the facility.

4.3.6 The visitors are advised to check-in and check-out when entering and leavening the facility. A visitor logbook is maintained, and security will scrutinize any visitor before entering the facility and connected to the specific person in charge of the visit.

4.3.7 The visitors are screened and verify the visitation before granting access to the facility as well as areas where a visitor will have specific access to.

4.3.8 The visitors are accepted through the general office and accompanied by a company executive during the entire visit. The visitor is not left alone for any circumstances other than the use of the washroom or inside the visitor's resting area.

4.3.9 The company shall provide passes for all the registered suppliers while security has been advised to scrutinize suppliers entering to the facility with specific persons registered.

4.3.10 If allergens are handled by the facility for any specific activity, they shall be segregated from the products or raw materials to avoid any cross-contamination.

4.4 Secure Facility Controls

4.4.1 All the perimeter areas are demarcated, high fenced, with specific in and out gates that are strictly controlled by the security personnel and surveillance camera (CCTV) system. Fencing, surveillance cameras, and security personnel shall be deployed as to the requirements and risks identified.

4.4.2 The CCTV surveillance is further extended to production areas while specific cameras shall be implemented to overlook rolling room operations with the clear view of rollers and vibrators to provide clear evidence that there are no any sugar adulterations are taking place, which is strictly prohibited. Bottles and cans are also strictly prohibited from taking into the rolling room.

4.4.3 The facility shall be kept enclosed all the time; the entrance gates are controlled by security personnel.

4.4.4 All the buildings or facilities shall be under lock and key when they are not in use. Storages, vehicles, containers, trailers, or any other temporary holding units which are not being used are kept locked.

4.4.5 Secure all the agricultural chemicals and verify inventory on a scheduled basis with restricted access keys.

4.4.6 All the specific keys to the doors shall be kept strictly controlled; there is a specific key registry, which identifies the specific key holders for each access door in the facility. The areas specifically restricted for general employees, visitors, and untrained employees will be demarcated, informed, and signboards are posted. If any employee is given specific access to a restricted area where it is not his/her duty station, they have to record, sign and receive the specific access keys, which will be strictly regulated and controlled on the basis of case by case.

4.4.7 When the employees are terminated or resigned, uniforms, name tags, identification badges shall be collected from them prior to termination/resignation.

4.4.8 The company mails are collected at the security point, verified, recorded, and delivered to the office; the mails are distributed through the administration officer, asking individually to collect them from the office.

4.4.9 Personnel items are not allowed out of the restroom; all the belongings of employees are kept away from the production and storage areas.

4.4.10 Computer and network access shall be granted to the specific personal according to their job functions once it is a necessity; other personnel is restricted to the access.

4.4.11 All the transactions are recorded, and the company has established a traceability system, which includes computer-based transactions and all the manual transactions.

4.4.12 The security personnel are advised to conduct routine security inspections of the premises for signs of tampering, criminal or terrorist activities, or any noticeable incidents.

4.4.13 Security has been instructed to verify the security of doors and windows while maintaining a given logbook and a checklist.

4.4.14 The facility has been equipped with an emergency lighting system and backup generator, which is automatically self-start and connected to the facility power system at a power failure.

4.4.15 Floor plans, product flow plans, or any segregation charts that are relevant to the facility, its production, or any other operations are up-to-date and available with responsible persons.

4.5 Secure Supply Controls

4.5.1 The company has established a procedure for supplier evaluation and acceptance (PRP 12) as well as a policy for rejecting deliveries or materials that do not comply with quality/safety inspections or are suspected of any sort of contamination or threat. Unauthorized deliveries are rejected and returned to the sending party.

4.5.2 The company has established a food safety procedure for withdrawal and recall (FSP 7), which is capable enough to manage an incident of withdrawal or recall where relevant persons are appointed with given responsibilities and proof tested with a recorded mock recall event every year.

4.5.3 The transportation department has authorized for pre-notification and scheduling of all shipments and empty trucks. "Holding" shipments that have not been pre-scheduled shall be kept at a predetermined location until the load can be verified.

4.5.4 Delivery schedules shall be established, unloading crews are supervised, and product or raw materials are being inspected before acceptance and send samples to the safety/quality checks if necessary. The materials suppliers and drivers are required to check in prior to offloading to verify shipping documentation and the driver's identity.

4.5.5 The company has established procedures for inspection of incoming shipments for damage and signs of tampering. The specific notices are issued requiring all doors and hatches on incoming shipments to be sealed with a numbered seal that corresponds with a seal number on the accompanying shipping documentation. The company also has advised the relevant personnel on sealing outgoing shipments and annotating the seal number on accompanying documents.

4.5.6 The company will not accept any returned/refused containers or reusable packing materials unless they are sanitized and certified for reuse. However, case by case approval shall be granted on such individual cases.

4.5.7 Products that are returned due to tampering issues are properly inspected, evaluated, reported, and recorded for any threats.

4.5.8 The company has established a system to recognize and report discrepancies in finished products, raw materials, packaging/labels, and chemical materials.

4.5.9 The food defence team shall keep an eye on the sugar stocks of the company, and continuous stock balances are updated, verified, and traced for the intended use. The sugar will be purchased in small quantities for employee welfare supplies which have to be strictly controlled and monitored with audited records. A checklist will be implemented to maintain welfare supplies.

4.5.10 Domestic products and imported products or items shall be stored separately.

4.6 Background Screening

4.6.1 The background screening shall be conducted through comprehensive reference checks with previous employers (two prior employers), which should include validation of employment dates and compensation details in addition to qualitative information.

4.6.2 All provided educational qualifications, including degrees, diplomas, or certificates, will be independently checked for management and executive staff. If employment dates have not been validated by the reference checks, they need to be separately or independently validated where gaps in employment should be thoroughly investigated.

4.6.3 All the employees specifically check for their work history with factories that are known to be adulterating black tea with materials such as sugar or ferrous compounds (most common intentional adulterations in the tea industry for economic reasons) in the past. Such employees (especially management staff candidates) will be immediately disqualified for employment in the company.

4.6.4 A police certificate shall be included in the personnel file, criminal records should be independently checked as the final step of the employment selection process. If it is not legally possible to obtain access to criminal records, a Certificate of Good Conduct from the police department or court jurisdiction will be pursued where it is possible and available.

4.6.5 Credit checks shall be conducted for specific roles that, by their nature, provide easy access to the funds of the Company or its employees, i.e., certain Finance/Payroll/HR/Senior management positions. Driving records shall be verified for those positions involved with driving automobiles, trucks, tractors, forklifts, etc.

4.6.6 The suppliers who can demonstrate their ability to carry out effective screening measures will be preferred over those who cannot. In accordance with local laws, the company expects suppliers to carry out a reasonable level of reference checking on their individual suppliers.

5.0 References

5.1 Company database on product recalls, complaints, and market returns
5.2 Company employee personnel files
5.3 Product identification and traceability - FSMS 02/FSP/04
5.4 Gap analysis/annual self-assessment checklist - FSMS 04/PRP/13/CL/01
5.5 HR manual
5.6 Hazard Assessment
5.7 Emergency preparedness and response - FSMS 02/FSP/05
5.8 Supplier evaluation and selection - FSMS 02/PRP/12

6.0 Records

6.1 Background screening reports
6.2 Externally available data on similar threats to the production
6.3 Visitor and employee access control records
6.4 Product or security complaints, analysis, and actions taken records
6.5 Self-assessment and annual evaluation records
6.6 System update records and new risk assessment conducted

ISO 22000 FOOD SAFETY MANAGEMENT SYSTEM SELF ASSESSMENT FORMAT FOOD DEFENSE SELF ASSESSMENT CHECKLIST			
Firm Name:		Plant Name:	
Address:			
Type of Audit:	Self-Assessment:	Date (s) Conducted:	
Food Defense Team:		Food Safety Team Leader:	General Manager:
Hazard Assessment:	Hazard Analysis Issue Date:	HACCP Plan Issue Date:	OPRP Plan Issue Date:
TACCP/VACCP Assessment:	TACCP/VACCP Analysis Issue Date:	TACCP Plan Issue Date:	VACCP Plan Issue Date:

Plan Management and Exercise - ALL FACILITIES	Yes	No	N/A	Comments
Are a designated site security team and Food Defense Coordinator appointed to implement and oversee the Security and Food Defense plan?				
Are annual Security and Food Defense self-assessments conducted? Does the facility quarterly assess the security of its overall posture through the performance of a self-inspection of Physical security systems and effectiveness of security procedures and protocols? Is the site security and food defense plan reviewed (and revised if necessary) at a minimum annually?				
Are the site security and food defense plan treated as proprietary company information?				
Is key emergency contact information for local fire, law enforcement, and ambulance maintained in the plan?				
Is the above contact information reviewed and updated annually?				
Are evacuation procedures in the defense plan?				
Are procedures in place to control access during an emergency to authorized personnel only?				
Are periodic drills conducted on the operational elements of the security plan?				
Food Defense Planning Considerations - Warehouses, Plants, and Sites that handle food Is key emergency contact information for public health regulators included in the plan?				
Is there an established liaison between the site and public health officials and appropriate security officials, and other Law Enforcement officials?				
Is there an established relationship between the facility and the appropriate analytical laboratories for possible assistance in the investigation of product tampering cases?				
Are procedures for responding to threats of product tampering included in the plan?				
Are procedures for responding to actual incidents of product tampering detailed in the plan?				
Are employees encouraged to report signs of possible product tampering?				
Does the facility have a documented recall plan? Are procedures in the recall plan reviewed and updated as necessary?				

Physical and Procedural Security – ALL FACILITIES	Yes	No	N/A	Comments
Are the grounds groomed and well maintained with clear lines of sight?				
Are the facility's exterior areas monitored for signs of a suspicious activity or entry (e.g., by guard or CCTV [Closed Circuit TV])?				
Is there sufficient outside lighting to allow detection of unusual activities on any part of the establishment outside premises during non-daylight hours?				
Do emergency exits have self-locking doors and/or alarms?				
Are all non-employees (e.g., outside vendors, contractors, or similar visitors) registered upon entry?				
Is positive identification (government-issued photo identification) required for outside vendors and delivery personnel, contractors, or similar types of visitors to verify identity when granting access to the facility?				
Are the following secured with locks, seals, or sensors at all times to prevent entry by unauthorized persons?				
Outside doors and gates Windows Roof Openings Vents Trailers, railcars, containers, etc. Utility controls Storage areas Is there emergency lighting within the facility?				
Is active surveillance of the inside facility and operations maintained through CCTV (Closed Circuit TV)? Are emergency systems tested periodically?				
Are the locations of controls for emergency systems clearly marked?				
Are access to sensitive storage, manufacturing, etc., areas restricted and access controlled to prevent entry by unauthorized personnel?				
Are visitors, guests, and other non-establishment employees (e.g., contractors, salespeople, truck drivers) restricted to non-production and storage areas unless accompanied by an authorized employee?				
Are evacuation routes posted?				
Are procedures in place to routinely check toilets, maintenance closets, personal lockers, and storage areas for concealed people or packages?				
Is a regular inventory of facility keys conducted?				
Are the central controls for the following restricted (e.g., by the locked door/gate or limiting access to designated employees) to prevent access by unauthorized persons?				
Heating, Ventilation, and A/C systems Propane Gas Water Systems Electricity Disinfection systems?				

Personnel Security Procedures - ALL FACILITIES	Yes	No	N/A	Comments
Are background checks conducted on prospective employees prior to hiring?				
Are background checks required on contract employees?				
Do employees receive training on security procedures as part of their orientation training?				
Are procedures in place to ensure positive identification/recognition of all employees?				
Are identification procedures in place to ensure the positive identification/recognition for temporary employees and contractors (including construction workers, cleaning crews, and truck drivers) in the facility?				
Are procedures in place to screen employees entering the establishment during business hours and non-business hours?				
Are procedures in place to screen the entry of contractors into the establishment during business hours and non-business hours?				
Is access to sensitive areas of the facility denied to those individuals performing work functions, such as temporary employees, construction workers, cleaning crews, truck drivers, and contractors?				
Are procedures in place to ensure clear identification of personnel with their specific functions?				
Does the facility have a policy regarding items (cameras etc.) that employees or visitors maybe bring into the facility?				
Is a policy in place on what personal items may and may not be allowed inside production, warehouse, and food storage areas?				
If employees are provided with lockers for personal items, are announced and unannounced inspections of employees' lockers conducted (within the limits of local law)?				
Are employees instructed on the proper protocol for handling and reporting tampering, illness, or other threats?				

Food Production – Plants and Sites that Handle Food	Yes	No	N/A	Comments
Is access to product production and preparation areas limited to employees, escorted visitors, and regulatory inspection personnel only?				
Is the mixing and batching of product and ingredients and other operations where large amounts of exposed product are handled continuously monitored?				
Are lines that handle and transfer products, water, oil, or other ingredients monitored to ensure integrity?				
Is the packaging integrity of ingredients examined for evidence of tampering before use?				
Are records maintained to ensure the capability to trace-back raw materials to suppliers?				
Are records maintained to ensure the capability to trace-forward finished products to vendors?				

Storage Security - Warehouses, Plants, and Sites that handle food	Yes	No	N/A	Comments
Is access to raw product storage areas, including holding coolers, restricted (e.g., by the locked door/gate) to designated employees?				
Is an access log maintained for raw product storage areas?				
Is access to ingredient storage areas restricted to designated employees only?				
Is an access log maintained for ingredient storage areas?				
Is access to finished product storage areas restricted to designated employees?				
Is access to external storage facilities restricted to designated employees only?				
Are periodic security inspections of storage facilities and warehouses (including temporary storage vehicles) conducted?				
Are records maintained on facility security inspection results?				
Are product labels and packaging held in a secure area to prevent theft and misuse?				
Is the inventory of finished products regularly checked for unexplained additions and withdrawals from existing stock?				
Is the access to inside and outside storage areas for hazardous materials/chemicals such as pesticides, industrial chemicals, cleaning materials, sanitizers, and disinfectants restricted to designated employees?				
Are hazardous material/chemical storage areas separated from production areas of establishment?				
Is a monthly inventory of hazardous materials/chemicals maintained?				
Are discrepancies in the inventory of hazardous materials/chemicals immediately investigated?				
Are the storage areas for hazardous materials/chemicals separated from stock, storage, or production/preparation areas?				
Is a procedure in place to receive and securely store hazardous chemicals?				
Is a procedure in place to control the disposition of hazardous chemicals?				
Is access to incoming water lines and water wells restricted? (e.g., by locked door/gate or limiting access to designated employees)				
Is access to storage tanks for potable water restricted? Are potable water lines periodically inspected for possible tampering?				

Laboratories – Sites with laboratories will complete this section.	Yes	No	N/A	Comments
Is access to laboratories restricted (e.g., by a locked door) to designated employees?				
Is an access log maintained for laboratories?				
Are periodic security inspections of laboratories conducted?				
Are records maintained on security inspection results?				
Is the inventory of controlled reagents regularly checked for unexplained additions and withdrawals from existing stock?				
Are procedures for reporting to incidents of missing reagents reviewed annually and updated as required?				

Shipping and Receiving - Warehouses, Plants, and Sites that handle food	Yes	No	N/A	Comments
Are trailers on the premises maintained under lock and/or seal when not being loaded or unloaded?				
Are deliveries monitored and controlled by the shipping and receiving personnel?				
Are outgoing shipments of the product sealed with tamper-evident seals?				
Are the seal numbers on outgoing shipments documented on the shipping documents?				
Are incoming delivery trucks inspected for unusual cargo or activity?				
Are drivers' identities and credentials verified against a valid government-issued driver's license?				
Is there a staging of incoming deliveries prior to entry into the facility?				
Are vendors of packaging materials and labels selected with consideration of food defense measures implemented by vendors?				
Is access to loading docks controlled to avoid unverified or unauthorized deliveries?				
Is advance notification from suppliers (by phone, e-mail, or fax) required for all incoming deliveries?				
Are suspicious alterations in the shipping documents immediately investigated?				
Are all deliveries verified against the roster of scheduled deliveries?				
Are unscheduled deliveries held outside facility premises pending verification?				
Are off-hour deliveries accepted?				
If off-hour deliveries are accepted, is prior notice of the delivery required?				
If off-hour deliveries are accepted, is the presence of an authorized individual to verify and receive the delivery required?				
Is the integrity of internal compartments in the trailer, lot packaging, or in-transit security checks for less-than-truckload (LTL) or partial load shipments of materials verified?				
Is the integrity of incoming shipments of raw products, ingredients, and finished products checked at receiving dock for evidence of tampering?				

REMARKS			
See attached assessment discussion sheet(s) for details/explanations on nonconformities identified			
NAME OF FDTL:		SIGNATURE:	
DATE:		SECURITY EXPERT:	

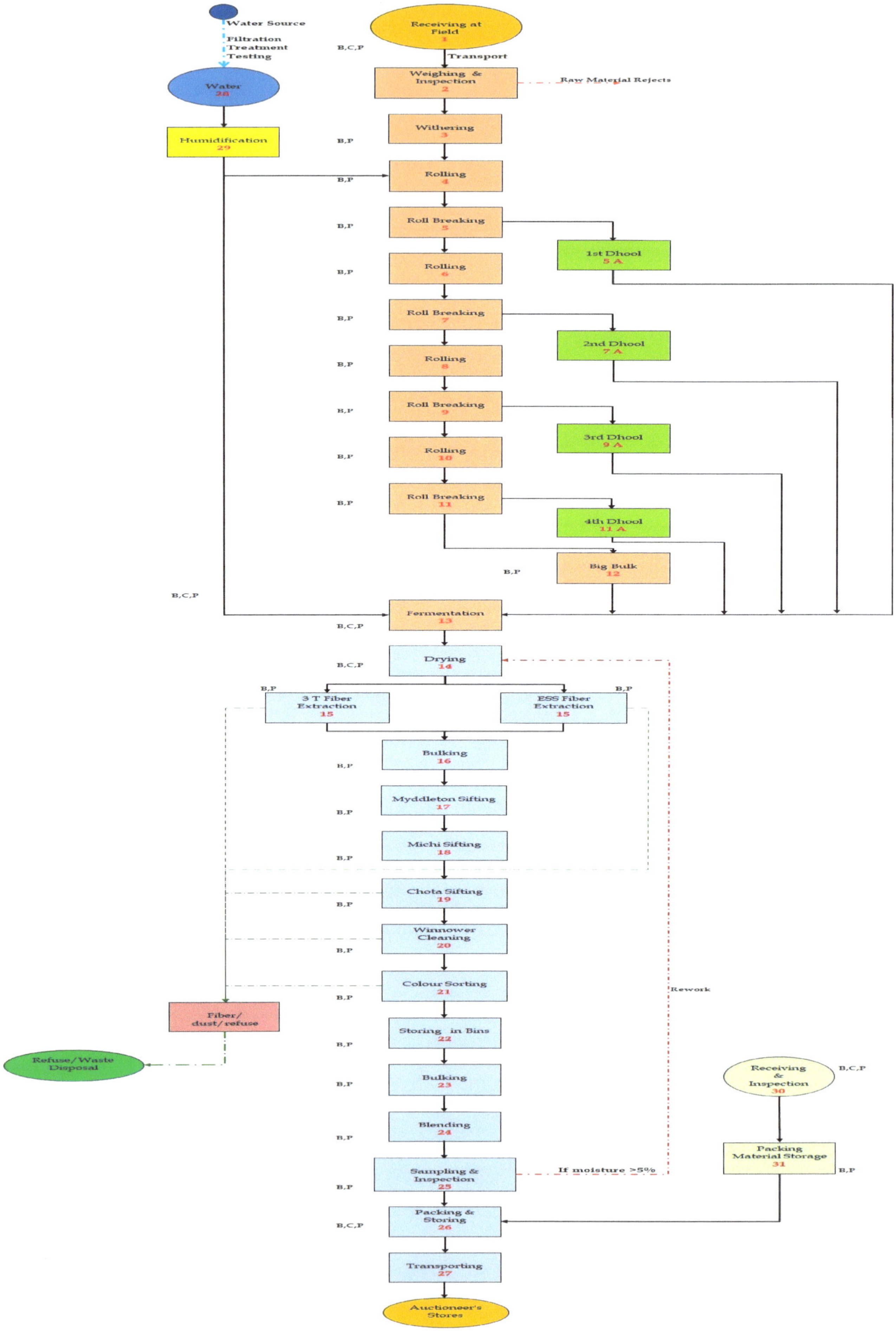
Water Source
Filtration
Treatment
Testing
Water
28
Humidification
29
Receiving at Field
1
Transport
Weighing & Inspection
2
Raw Material Rejects
Withering
3
Rolling
4
Roll Breaking
5
1st Dhool
5 A
Rolling
6
Roll Breaking
7
2nd Dhool
7 A
Rolling
8
Roll Breaking
9
3rd Dhool
9 A
Rolling
10
Roll Breaking
11
4th Dhool
11 A
Big Bulk
12
Fermentation
13
Drying
14
3 T Fiber Extraction
15
ESS Fiber Extraction
15
Bulking
16
Myddleton Sifting
17
Michi Sifting
18
Chota Sifting
19
Winnower Cleaning
20
Colour Sorting
21
Storing in Bins
22
Bulking
23
Blending
24
Sampling & Inspection
25
If moisture >5%
Rework
Packing & Storing
26
Transporting
27
Auctioneer's Stores
Fiber/ dust/ refuse
Refuse/Waste Disposal
Receiving & Inspection
30
Packing Material Storage
31
B,C,P
B,P

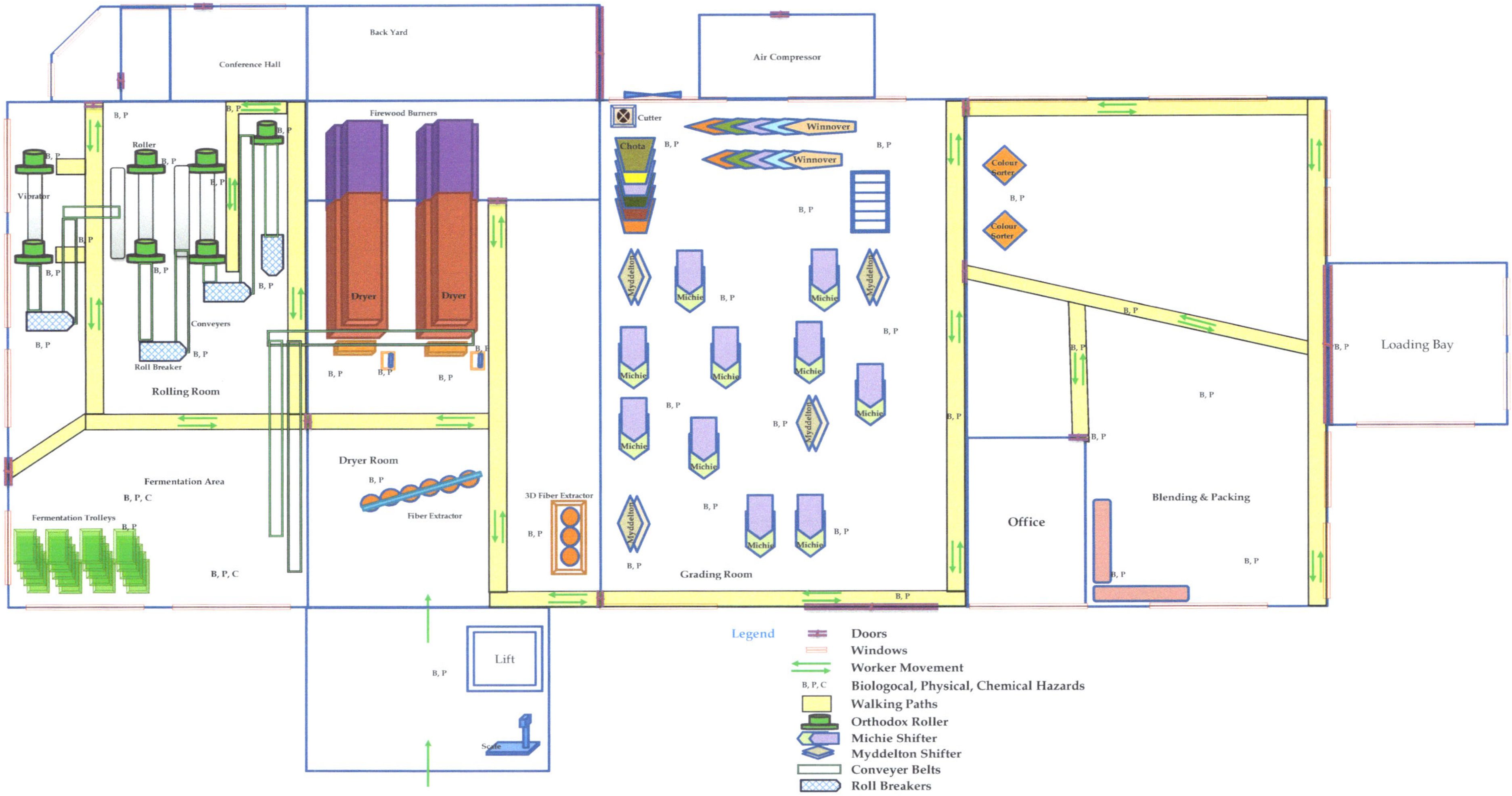

Conference Hall
Back Yard
Air Compressor
Firewood Burners
Cutter
Chota
Winnover
Winnover
Colour Sorter
Colour Sorter
Roller
Vibrator
Conveyers
Roll Breaker
Rolling Room
Dryer
Dryer
Myddelton
Michie
Dryer Room
Fiber Extractor
3D Fiber Extractor
Fermentation Area
B, P, C
Fermentation Trolleys
Grading Room
Office
Blending & Packing
Loading Bay
Lift
Scale
B, P
Legend
Doors
Windows
Worker Movement
Biological, Physical, Chemical Hazards
Walking Paths
Orthodox Roller
Michie Shifter
Myddelton Shifter
Conveyer Belts
Roll Breakers

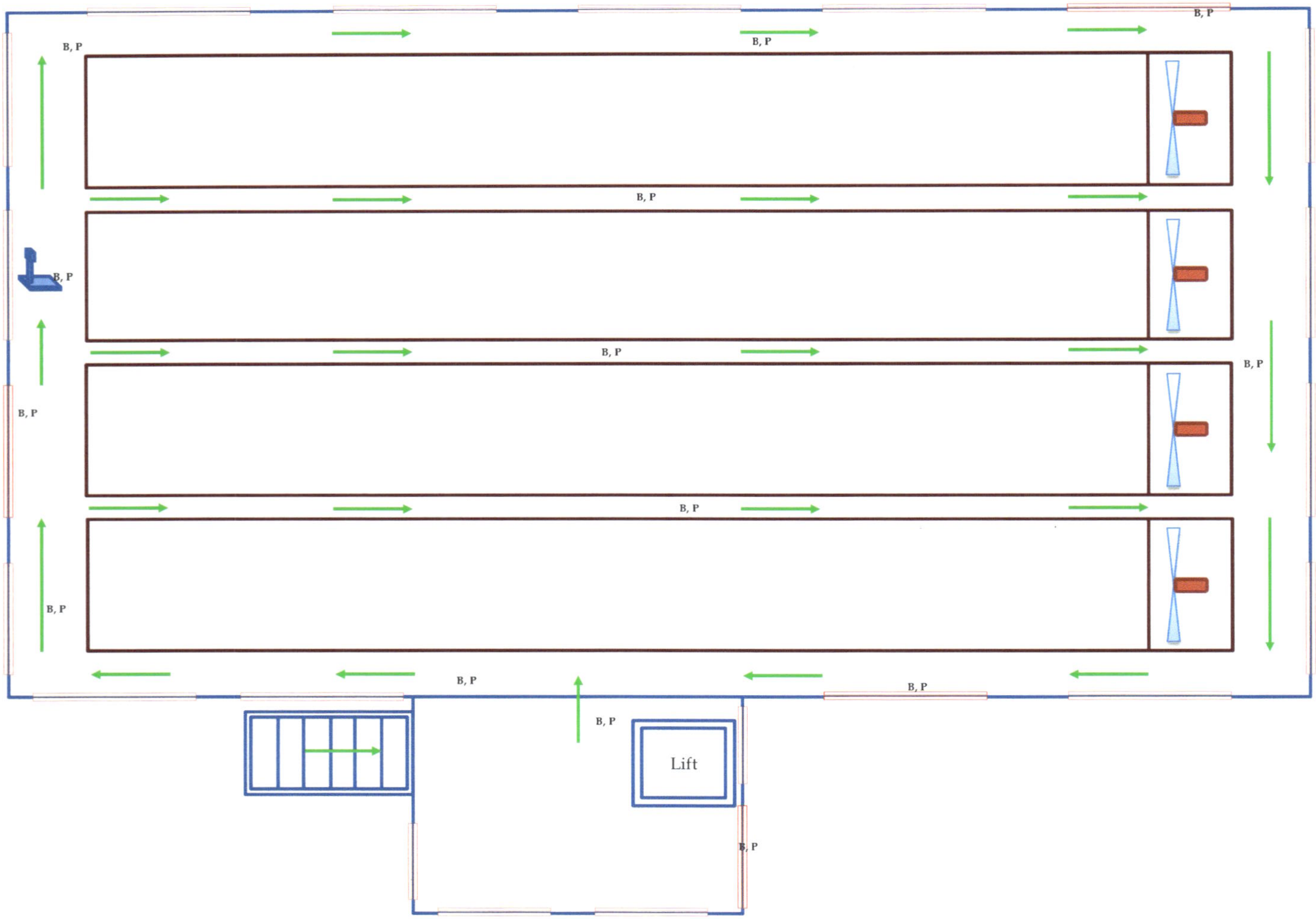
B, P
B, P
B, P
B, P
B, P
B, P
B, P
B, P
B, P
B, P
B, P
B, P
B, P
B, P
Lift

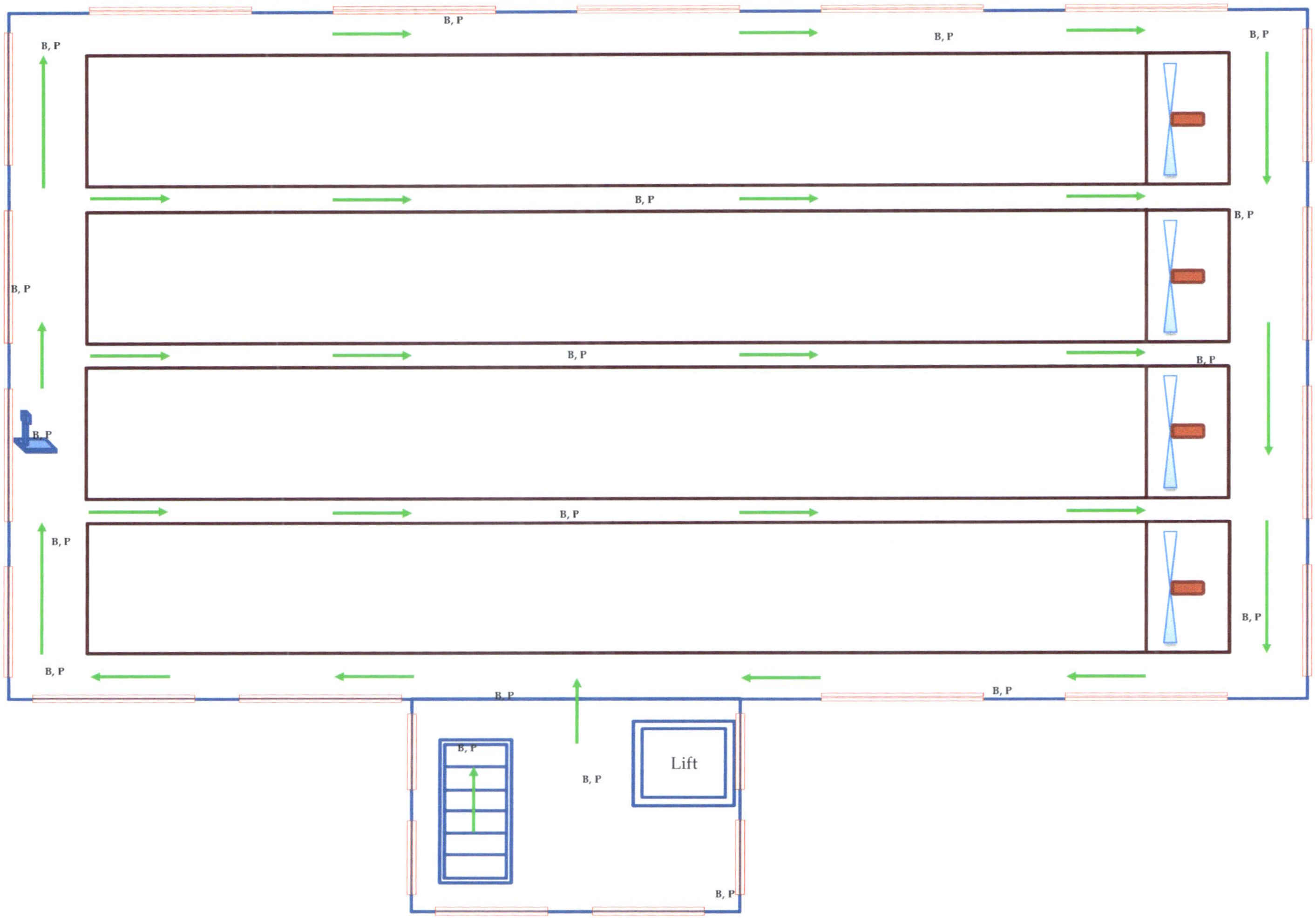
Lift

Raw material/ Ingredients/ Product Contact Material Description	
Green Leaf	
Name	Green Leaf/*Camellia sinensis (L) O. Kuntze*
Biological, Chemical, Physical characteristics	Tender shoots having bud and 2 -3 leave that are free of physical damages and contamination.
Composition or formulated ingredients, including additives and processing aids	Polyphenolic compounds, caffeine, amino acids, carbohydrates, chlorophyll, cells, fibre, and water
Source	Tea leaves – vegetative/plant leaves
Origin	Sri Lanka – Low grown areas, mainly from Matara, Ratnapura, and Galle districts
Method of production	Manual plucking every 7days cycle from properly managed plantations owned by mostly smallholders
Packaging and delivery method	Loosely packed in plastic crates or gunny sacks and deliver to the factory through ventilated trucks.
Storage conditions and shelf life	Store at ambient temperature, and shelf life is about 3 days.
Preparation and/or handling before use of processing	Directly used
Food safety related acceptance criteria or specifications	Conform to SLS 1315 part I good agricultural practices for the cultivation of tea.

Raw material/ Ingredients/ Product Contact Material Description	
Paper Sacks	
Name	Multilayer paper sack/ Kraft liner board sacks
Biological, Chemical, Physical characteristics	Nontoxic, non-recycled food-grade raw materials (softwoods) are used, and chemicals are applied to bleach the pulp.
Composition or formulated ingredients, including additives and processing aids	Natural hard and rough paper sacks including an inner aluminium foil layer as a barrier for moisture and air. Composition: Outer ply (Kraft liner board) min 270gsm; Inner ply (Kraft paper) min 60gsm; Polyethylene min 20gsm; Aluminium/ Metalized polyester min 19gsm
Source	Vegetative material - plant stem/leaves
Origin	Sri Lanka/India
Method of production	Wood pulp for sack paper is made from softwood by the Kraft process. The long fibres provide the paper its strength, and wet strength chemicals are added to even further improve the strength. Both white and brown grades are made. Sack paper is then produced on a paper machine from the wood pulp.
Packaging and delivery method	Tightly packed in bundles which are wrapped in paper sheets and deliver to the factory through fully closed trucks.
Storage conditions and shelf life	Store at cool dry cleaned the place and shelf life is around 4 years.
Preparation and / or handling before use of processing	Directly used
Food safety related acceptance criteria or specifications	Conform to SLS packaging standards and ISO 9001, GMP.

Raw material/ Ingredients/ Product Contact Material Description	
Potable Water	
Name	Potable Water
Biological, Chemical, Physical characteristics	A colourless liquid without any odor or flavours.
Composition or formulated ingredients, including additives and processing aids	100% water (H_2O)
Source	H_2O/Mineral
Origin	Deniyaya - Sri Lanka/ taken out of a hilly stream at Panilkanda that is properly fenced and secured against contamination from wild animals or humans.
Method of production	Purified using a set of standard filters and Ultra Violet Light for the elimination of pathogenic bacteria and any particles that remain in the stream.
Packaging and delivery method	Deliver through PVC pipeline system
Storage conditions and shelf life	Store closed plastic tanks after purification
Preparation and/or handling before use of processing	Directly used
Food safety related acceptance criteria or specifications	Conform to SLS 614 potable water

Raw material/ Ingredients/ Product Contact Material Description	
Roller Table	
Name	Central Fitting (Cone) Table Fitting (Batten)
Biological, Chemical, Physical characteristics	Dark Brown Hard Appearance. High strength & Resistance to Termites and Pinhole Borers. Characteristic flavors are not imparting to the material in contact with wood.
Composition or formulated ingredients, including additives and processing aids	Made out of "Kithul" (Cariyota Uranus) wood. Chemical Composition : Cellulose, Hemicellulose, Lignin, high strength fibre with no activity.
Source	Vegetative material - plant stem
Origin	Sri Lanka
Method of production	Wood planks are shaped and smoothen according to the surface roughness and twisting action requirements which connected perfectly on top of the table. Internal and external planning and sanding of the wood give the finishing touches. Finally the relative accessories are installed depending on roller requirements.
Packaging and delivery method	The Wood planks are delivered to the factory and carpenters build the table tops at roller manufacturing premises.
Storage conditions and shelf life	Infinite Shelf life; however it is used around two years. Tear and wear of the surfaces due to everyday rolling needs refurbishing time to time.
Preparation and/or handling before use of processing	Cleaning with water and Detergent to reduce the possibility of unwanted bacteria growth.
Food safety related acceptance criteria or specifications	None

Raw material/ Ingredients/ Product Contact Material Description	
Stainless Steel	
Name	Stainless Steel - Grade 304 Stainless Steel - Grade 316/316L
Biological, Chemical, Physical characteristics	Shiny Metallic Appearance. Corrosion Resistance. 304 SS Maximum exposure temperature is about 1500°F (816°C), Good oxidation resistance in intermittent service to 870°C and in continuous service to 925°C, & Non-hardenable by heat treatment. 316/316L SS Good oxidation resistance in intermittent service to 870°C and in continuous service to 925°C. Higher resistance to pitting and crevice corrosion is required in chloride environments than 304 SS
Composition or formulated ingredients, including additives and processing aids	304 SS Fe, <0.08% C, 17.5-20% Cr, 8-11% Ni, <2% Mn, <1% Si, <0.045% P, <0.03% S 316/316L SS Fe, <0.03% C, 16-18.5% Cr, 10-14% Ni, 2-3% Mo, <2% Mn, <1% Si, <0.045% P, <0.03% S
Source	Metal/inorganic mineral compounds
Origin	China/ India
Method of production	Iron ore, chromium, silicon, nickel are melted together in an electric furnace for about 8 to 12 hours, and then the mixture is cast into one of several shapes, then it is cast into a solid form. Later on, the steel is heat-treated and then cleaned and polished to give it the desired finish. Finally, it is packaged and sent to manufacturers, who weld and join the steel to produce the desired shapes.
Packaging and delivery method	Imported from China – International Shipping
Storage conditions and shelf life	Excellent in a wide range of atmospheric environments and many corrosive media. Subject to pitting and crevice corrosion in warm chloride environments and to stress corrosion cracking above about 60°C.
Preparation and/or handling before use of processing	Cleaning with Water and Detergent prior to the production process
Food safety related acceptance criteria or specifications	ISO Standards ICS67.260: Plants and equipment for the food industry EN, ASTM, and JIS standards

End Product Description	
Made Tea	
Product Name	Orthodox Black Tea (Main Grades & Off Grades)
Composition	5 – 7% of moisture and the rest is dried tea leaves.
Biological Physical and Chemical Characteristics	High temperature long time (HTLT) treated in an oven and stored at ambient temperature 28°C, Contains no preservatives which must comply with ISO 3720 for black tea on chemical, biological and physical requirements. Water Extract, min 32% (m/m); Total Ash, min 4%-max 8%; Water Soluble Ash of total ash, min 45%; Alkalinity of water – Soluble Ash (as KOH), min 1%-max3% (m/m); Acid Insoluble Ash, max 1% (m/m); Crude Fibre max 16.5%. Foreign matter: completely free Heavy metals: Iron – max 500mg/kg; Copper – max 100mg/kg; Lead – max 2mg/kg; Zinc – max 100mg/kg; Cadmium – max 0.2mg/kg; Microbiological Requirements: APC – max 10,000cfu/g; Yeast & Mould – max 1000cfu/g; Total Coliform – max 10 MPN/g; E. coli – absent/g; Salmonella – absent/25g; Pesticide Residues: 27 specific pesticides listed according to the country of export by SLTB "basic requirements for Black Tea: ISO 3720:1986/ Corrigendum 1:1992(E) & 2:2004 (E) and/ or SLS 135:1979 (UDC 663.951) /SLTB Ref: AD (Lab). SLTB – MRLs(A): April 2010*
Intended shelf life	2 Year
Storage Conditions	A cold, dry place with less RH at ambient temperature.
Packaging materials	Food grade Multilayer Craft Paper sacks with Aluminium foiled inside to prevent moisture absorption. Sacks are containing 50Kg, 28Kg, 38Kg, 40Kg, 42Kg, or any custom size of made tea.
Labelling relating to food safety and/or instructions for handling and usage	Store at a cold, dry place away from wet conditions. Handle with care to minimize the damages to paper sacks.
Methods of distribution	Transport up to auctioneer's stores in the company closed container truck.
Intended use	After adding to the boiling water for 5 minutes, direct consumption by the general public, including high-risk groups.

* *Please attach the SLTB circular to this document*

HAZARD IDENTIFICATION - PROCESSING STEPS

	Process Step Raw material/PCM/ Ingredient Site Schematics End products	Hazard Identified	Acceptable Levels (Limits)	Controlled at
1	Leaf collection at field	B – Introduction & proliferation of microbes at the field B – Insects or insect parts during leaf plucking, transport, or collection C – Pesticide or herbicide residues P – Non-tea leaves during leaf plucking P – Foreign matters (wood, metal, plastic dust, soil)	Temperature at receiving 28 ±2°C Supply of acceptable air circulation at transportation Quality & safety assurance while leaf plucking Pesticide or herbicide guidelines according to the chemical for harvesting time regulations No any other leaves or foreign matters Creates, gunny/poly sacks, transporting vehicles at proper hygienic conditions at receiving Visual observation (company specification)	PRP 12 – Supplier Evaluation & Selection Supplier Quality Assurance Supplier education (training) & practices Cleanliness of utensils or bins Prerequisites at plantation Good Agricultural Practises PRP 05 – Storing, handling & transportation of material PRP 02 – Management of purchased material
2	Brought leaf receiving & inspection	**B** – Infestation of insects to harvested leaves due to cross-contamination **C** – Pesticide or herbicide residues due to inadequate practices **C** – Chemical changes due to high temperature & sunlight P – Foreign matters (wood, metal, plastic dust, soil) P – Matured leaves or stems P – Compression & impact damage	No extraneous matters or pests/excreta at inspection Herbicides – after 14 days Pesticides/Fungicides – after 21 days Temperature at receiving 28 ±2°C RH at receiving 80 ±5 % Leaves at uniform fresh conditions & 75% "Ran Dalu" verification Visual observation (company specification)	PRP 12 – Supplier Evaluation & Selection Supplier Quality Assurance Supplier education (training) & practices Collection with a time gap PRP 02 – Management of purchased material PRP 05 – Storing, handling & transportation of material

	Process Step	Hazard Identified	Acceptable Levels (Limits)	Controlled at
3	Withering	B – Infestation of insects P – Foreign matters (wood, metal, plastic & dust) P – unsystematic withering (less/over)	No insects or flies can penetrate to the withering area Non at visual observation Appropriate withering at RH 80% & 3°C difference for dry bulb & wet bulb temp.	PRP 01 – Construction, layout, and equipment PRP 03 – Preventive maintenance PRP 05 – Storing, handling & transportation of material PRP 07 – Pest control PRP 09 – Supply of air water & energy WI – Withering Area
4 - 12	Rolling	B – Introduction & proliferation of microbes due to increase in temperature & RH, hygienic condition of the floor C – Lubricants (oil & Grease) P – Foreign matters (wood, metal, plastic, soil) due to inappropriate leaf handling	Maintenance of proper hygienic conditions inside the Rolling room creates & floor RH 85% & temperature 26 °C Use of food-grade lubricants Proper monitoring of leaf handling	PRP 06 – Personal Hygiene PRP 03 – Preventive maintenance PRP 10 – Work environment PRP 11 – Cleaning & sanitation of plant machinery, equipment & utensils WI – Rolling Room
13	Fermentation	B – Introduction & proliferation of microbes due to increase in temperature & reduced RH, hygienic condition of the floor & RR P – Foreign matters (wood, metal, plastic dust, soil) due to inappropriate dhool handling	Maintenance of proper hygienic conditions inside the Rolling room creates & floor RH 85% & temperature 26 °C Proper monitoring of dhool handling	Control of RH Control of temperature Control of fermentation time PRP 06 – Personal Hygiene PRP 10 – Work environment PRP 11 – Cleaning & sanitation of plant machinery, equipment & utensils WI – Fermentation

	Process Step	Hazard Identified	Acceptable Levels (Limits)	Controlled at
14	Drying/Firing	B – Inappropriate (less) drying – survival of bacteria & spores C – Deposition of chemical residues from fuel-burning P – Extraneous matters (Ash, wood, metal, plastic & dust)	Kiln thermometer reading at 124 ±2°C (operational limit) 0% use of fuel/kerosene for kiln firing No ash or half-burned wood or waste in the kiln's air ducts at visual observation	Metal detection & collection PRP 01 – Construction, layout, and equipment PRP 03 – Preventive maintenance PRP 09 – Supply of air water & energy WI – Firing Room
15	Fibre extraction	P – Extraneous matters – especially metal due to the heavy mass B – Cross-contamination from operators P – Increase in moisture	0% Iron or metal at the final product Less than 7% moisture in the final product (operational limit)	Metal detection & collection PRP 06 – Personal Hygiene PRP 10 – Work environment PRP 11 – Cleaning & sanitation of plant machinery, equipment & utensils WI – Firing Room
16	Bulking	B – Cross-contamination from operators, equipment, floor & utensils P – Extraneous matters (wood, metal, plastic & dust) P – Increase in moisture	Maintenance of proper hygienic conditions inside the firing room creates & floor Non at visual observation Less than 7% moisture in the final product (operational limit)	PRP 06 – Personal Hygiene PRP 10 – Work environment PRP 11 – Cleaning & sanitation of plant machinery, equipment & utensils WI – Firing Room
17	Myddleton Sifting	B – Cross-contamination from operators, floor & utensils P – Extraneous matters (wood, metal, plastic & dust) P – Increase in moisture	Maintenance of proper hygienic conditions inside the grading room creates & floor Non at visual observation Less than 7% moisture in the final product (operational limit)	PRP 06 – Personal Hygiene PRP 03 – Preventive maintenance PRP 10 – Work environment PRP 11 – Cleaning & sanitation of plant machinery, equipment & utensils WI – Grading Room

	Process Step	Hazard Identified	Acceptable Levels (Limits)	Controlled at
18	Michi Sifting	B – Cross-contamination from operators, floor & utensils P – Extraneous matters (wood, metal, plastic & dust) P – Increase in moisture	Maintenance of proper hygienic conditions inside the grading room creates & floor Non at visual observation Less than 7% moisture in the final product (operational limit)	PRP 06 – Personal Hygiene PRP 03 – Preventive maintenance PRP 10 – Work environment PRP 11 – Cleaning & sanitation of plant machinery, equipment & utensil WI – Grading Room
19	Chota Sifting	B – Cross-contamination from operators, floor & utensils P – Extraneous matters (wood, metal, plastic & dust) P – Increase in moisture	Maintenance of proper hygienic conditions inside the grading room creates & floor Non at visual observation Less than 7% moisture in the final product (operational limit)	PRP 06 – Personal Hygiene PRP 03 – Preventive maintenance PRP 10 – Work environment PRP 11 – Cleaning & sanitation of plant machinery, equipment & utensils WI – Grading Room
20	Winnowing	B – Cross-contamination from operators, floor & utensils P – Extraneous matters (wood, metal, plastic & dust) P – Increase in moisture	Maintenance of proper hygienic conditions inside the grading room creates & floor Non at visual observation Less than 7% moisture in the final product (operational limit)	Metal detection & collection PRP 06 – Personal Hygiene PRP 03 – Preventive maintenance PRP 10 – Work environment PRP 11 – Cleaning & sanitation of plant machinery, equipment & utensils WI – Grading Room
21	Colour Sorting	B – Cross-contamination from operators, floor & utensils P – Extraneous matters (wood, metal, plastic & dust) P – Increase in moisture	Maintenance of proper hygienic conditions inside the grading room creates & floor Non at visual observation Less than 7% moisture in the final product (operational limit)	Metal detection & collection PRP 06 – Personal Hygiene PRP 03 – Preventive maintenance PRP 10 – Work environment PRP 11 – Cleaning & sanitation of plant machinery, equipment & utensils WI – Grading Room

	Process Step	Hazard Identified	Acceptable levels (Limits)	Controlled at
22 – 23	Storing & Bulking	B – Cross-contamination from operators, floor & utensils P – Extraneous matters (wood, metal, plastic & dust) P – Increase in moisture	Maintenance of proper hygienic conditions inside the grading room creates & floor Non at visual observation Less than 7% moisture in the final product (operational limit)	PRP 06 – Personal Hygiene PRP 03 – Preventive maintenance PRP 10 – Work environment PRP 11 – Cleaning & sanitation of plant machinery, equipment & utensils WI – Grading Room
24	Blending	B – Cross-contamination from operators, floor & utensils P – Extraneous matters (wood, metal, plastic & dust) P – Increase in moisture	Maintenance of proper hygienic conditions inside the grading room creates & floor Non at visual observation Less than 7% moisture in the final product (operational limit)	PRP 06 – Personal Hygiene PRP 03 – Preventive maintenance PRP 10 – Work environment WI – Grading Room
25	Sampling & Inspection	B – Cross-contamination from operators, floor & utensils P – Increase in moisture	Maintenance of proper hygienic conditions at sampling & inspections Non at visual observation Less than 7% moisture in the final product (operational limit)	PRP 06 – Personal Hygiene PRP 03 – Preventive maintenance PRP 10 – Work environment
26	Packing & Storing	B – Introduction & proliferation of microbes due to increase in RH, hygienic condition of the floor & SR P – Extraneous matters (dust, packing materials) P – pellet formation or granulation	Maintenance of proper hygienic conditions inside the packing room & storage, reduced RH levels 50 ±5%, 28 ±2°C room temperature	Control of RH Control of temperature PRP 05 – Storing, handling & transportation of material PRP 06 – Personal Hygiene PRP 07 – Pest control PRP 10 – Work environment PRP 11 – Cleaning & sanitation of plant machinery, equipment & utensils WI – Sifting & Packing

	Process Step	Hazard Identified	Acceptable Levels (Limits)	Controlled at
27	Transportation	P – Extraneous matters P – Increase of moisture	Vehicles shall not be used for transportation of any hazardous materials Use of fully closed vehicles for transportation	PRP 03 – Preventive maintenance PRP 05 – Storing, handling & transportation of material WI – Transportation Hygiene

HAZARD IDENTIFICATION – INGREDIENT

	Ingredients	Hazard Identified	Acceptable Levels (Limits)	Controlled at
28 – 29	Raw Water	B – Water not meeting the drinking water criteria established by SLSI (Microbial) C – Heavy metals C – Pesticides P – Hazardous extraneous material (metal)	SLSI Microbiological acceptable levels for Water (arithmetic values) SLSI H. Metal/Pesticide acceptable levels for Water (arithmetic values)	PRP 09 – Supply of air water & energy PRP 11 – Cleaning & sanitation of plant machinery, equipment & utensils WI – Water Purification

HAZARD IDENTIFICATION – PACKING MATERIALS

	Packing Materials	Hazard Identified	Acceptable Levels (Limits)	Controlled at
30 – 31	Paper Sacks	B – Introduction & proliferation of microbes due to improper cleaning B – Cobwebs & insects P – Extraneous matters (Dust)	0% Visual observation (company specification)	PRP 07 – Pest control PRP 10 – Work environment PRP 11 – Cleaning & sanitation of plant machinery, equipment & utensils

HAZARD IDENTIFICATION – RAW MATERIAL/ PRODUCT CONTACT MATERIAL/ INGREDIENT

	Raw Material/PCM	Hazard Identified	Acceptable Levels (Limits)	Controlled at
32	Green Leaves	B – Introduction & proliferation of microbes at the field B – Insects or insect parts during leaf plucking, transport, or collection C – Pesticide or herbicide residues P – Non-tea leaves while leaf plucking P – Foreign matters (wood, metal, plastic dust, soil)	Temperature at receiving 28 ±2°C Supply of acceptable air circulation at transportation Quality & safety assurance while leaf plucking Pesticide or herbicide guidelines according to the chemical for harvesting time regulations No any other leaves or foreign matters Crates, gunny/poly sacks, transporting vehicles at proper hygienic conditions at receiving Visual observation (company specification)	PRP 12 – Supplier Evaluation & Selection Supplier Quality Assurance Supplier education (training) & practices Cleanliness of utensils or bins Prerequisites at plantation (Good Agricultural Practises) PRP 05 – Storing, handling & transportation of material PRP 02 – Management of purchased material
33	Troughs /Rollers/ Conveyers/ Sifting Equipments/oven	B – Introduction & proliferation of microbes due to improper cleaning B – Cob webs & insects P – Extraneous matters (Ash, wood, metal, plastic & dust)	Visual observation (company specification)	PRP 07 – Pest control PRP 10 – Work environment PRP 11 – Cleaning & sanitation of plant machinery, equipment & utensils
34	Creates	B – Introduction & proliferation of microbes due to improper cleaning B – Cobwebs & insects P – Extraneous matters (Ash, wood, metal, plastic & dust)	0% Visual observation (company specification)	PRP 07 – Pest control PRP 10 – Work environment PRP 11 – Cleaning & sanitation of plant machinery, equipment & utensils

HAZARD IDENTIFICATION – SITE SCHEMATICS

	Site Schematics	Hazard Identified	Acceptable levels (Limits)	Controlled at
35	Main door at ground floor, three entrances & windows at rolling, packing & sifting, are directly opening to the outside environment	B – Introduction & proliferation of microbes due to cross-contamination through employee movements & airflow movements B – Infestation of insects or birds/rodents through openings P – Foreign matters (dust)	Limited movements of employees through finished products to raw products areas (not vice versa) & self-closing doors Netted windows with $5mm^2$ cleanable nets Visual observation (company specification) No openings without barriers to prevent insect, birds/rodents entrance	PRP 01 – Construction, layout, and equipment PRP 03 – Preventive maintenance PRP 06 – Personal Hygiene PRP 07 – Pest control PRP 09 – Supply of air water & energy PRP 10 – Work environment WI – RR/FR/SR/PR
36	The loft area is made of wooden structures & granite; high roof, windows & openings	B – the proliferation of microbes due to wooden/granite structures B – Infestation of insects or birds/rodents through openings P – Cobwebs	Proper painting with food-grade paints Management of hygienic conditions Visual observation (company specification) Preventive maintenance	PRP 01 – Construction, layout, and equipment PRP 03 – Preventive maintenance PRP 07 – Pest control PRP 10 – Work environment PRP 11 – Cleaning & sanitation of plant machinery, equipment & utensils
37	Firewood store	B – Infestation of insects or rodents & proliferation or breeding sites	Clean and systematic without chips/ muddy soil at visual observation	Practise of FIFO at firewood receiving & usage PRP 03 – Preventive maintenance PRP 07 – Pest control PRP 10 – Work environment WI – Firewood Store

HAZARD IDENTIFICATION – END PRODUCT

	End Product	Hazard Identified	Acceptable Levels (Limits)	Controlled at
38	Made Tea	B – Introduction & proliferation of microbes due to increase in RH, hygienic condition of the packaging P – Extraneous matters (dust, packing materials) P – pellet formation or granulation	Maintenance of proper hygienic conditions inside the storage, reduced RH levels 50 ±5%, 26 ±2°C room temperature Proper packaging	Control of RH Control of temperature PRP 05 – Storing, handling & transportation of material PRP 06 – Personal Hygiene PRP 07 – Pest control PRP 10 – Work environment PRP 11 – Cleaning & sanitation of plant machinery, equipment & utensils WI – End Product Handling

HAZARD CONTROL PLAN – EXTERNALLY CONTROLLED HAZARDS

SUPPLIER CONTROLLED HAZARDS	
PRODUCT: ORTHODOX BLACK TEA	
Hazards	Controlled by
Step 0: Leaf quality & safety at field collection	
B – Introduction & proliferation of microbes at the field B – Insects or insect parts while leaf fluking, transport, or collection C – Pesticide or herbicide residues P – Non-tea leaves during leaf plucking P – Foreign matters (wood, metal, plastic dust, soil)	Supplier education and practices; Supplier quality assurance Supplier education (training) & practices Cleanliness of utensils or bins Prerequisites at plantation Good Agricultural Practises – GAP
Step 1: Brought leaf receiving & inspection	
B – Infestation of insects to harvested leaves due to cross-contamination **C** – Pesticide or herbicide residues due to inadequate practices **C** – Chemical changes due to high temperature & sunlight P – Foreign matters (wood, metal, plastic dust, soil) P – Matured leaves or stems	Supplier quality assurance Supplier education (training) & practices Rejection deduction(%) of low-quality green leaves Advice for better practices

HAZARD ANALYSIS – ORTHODOX BLACK TEA

Ingredient/ PCM/ Processing Step	Potential Hazard (B, C, P)		Is the identified hazard of sufficient probability and severity to warrant control in HACCP plan?	Justification of Decision (If no, identify the prerequisite program that controls the hazard)	Control Measures (Significant hazards only)
Leaf collection at field 1	B	Growth of *Salmonella,* *E. Coli* & *Staphylococcus aureus* *Aspergillus* Insects, parts or excreta	*Salmonella* Risk S(3) * P(1) = 03 Low *E. Coli* S(3) * P(1) = 03 Low *Staphylococcus aureus* S(3) * P(1) = 03 Low S(3) * P(1) = 03 Low S(2) * P(1) = 02 Low	Supplier quality assurance programs & training GAP PRP 05 – Storing, handling & transportation of material PRP 02 – Management of purchased material	
	C	Pesticide or herbicide residues	S(3) * P(2) = 06 High	Supplier education & Herbicides – green leaf accepts after 14 days Pesticides/Fungicides – after 21 days	
	P	Foreign matters	S(1) * P(1) = 01 Low	Visual inspection at receiving	
Brought leaf receiving & inspection 2	P	Matured leaves or stems Compression & impact damage	S(1) * P(1) = 01 Low S(1) * P(3) = 03 Low	Percentage reductions Advice or rejection after inspection	
Withering 3	B	Infestation of insects	S(2) * P(2) = 04 Medium	PRP 03 – Preventive maintenance PRP 05 – Storing, handling & transportation of material PRP 07 – Pest control PRP 09 – Supply of air water & energy	
	P	Foreign matters (wood, metal, plastic & dust)	S(1) * P(3) = 03 Low		

Ingredient/ PCM/ Processing Step	Potential Hazard (B, C, P)		Is the identified hazard of sufficient probability and severity to warrant control in HACCP plan?	Justification of Decision (If no, identify the prerequisite program that controls the hazard)	Control Measures (Significant hazards only)
Rolling 4 – 12	B	Growth of *Salmonella,* *E. Coli* & *Staphylococcus aureus* *Aspergillus*	*Salmonella* Risk S(3) * P(1) = 03 Low *E. Coli* S(3) * P(1) = 03 Low *Staphylococcus aureus* S(3) * P(1) = 03 Low *Aspergillus* S(3) * P(1) = 01 Low	PRP 03 – Preventive maintenance PRP 06 – Personal Hygiene PRP 10 – Work environment PRP 11 – Cleaning & sanitation of plant machinery, equipment & utensils	
	C	Lubricants (oil & Grease)	S(2) * P(2) = 04 Medium		
	P	Foreign matters	S(1) * P(3) = 03 Low		
Fermentation 13	B	Growth & proliferation of *Salmonella,* *E. Coli* & *Staphylococcus aureus* *Aspergillus*	*Salmonella* Risk S(3) * P(2) = 06 High *E. Coli* S(3) * P(1) = 03 Low *Staphylococcus aureus* S(3) * P(3) = 09 High *Aspergillus* S(3) * P(2) = 06 High	EEC (European Economic Commission) Directive 493 describes as *Salmonella, E. Coli* & *Staphylococcus aureus* are Severe Pathogens.	Time control – within 2.5 hours process will be completed Control RH 20 ±5° C or 80 ±5° F RH 85°±5°
	P	Foreign matters	S(1) * P(3) = 03 Low	PRP 03 – Preventive maintenance	
Drying/Firing 14	B	survival of *Salmonella,* *E. Coli* & *Staphylococcus aureus* *Aspergillus*	*Salmonella* Risk S(3) * P(2) = 06 High *E. Coli* S(3) * P(1) = 03 Low *Staphylococcus aureus* S(3) * P(3) = 09 High *Aspergillus* S(3) * P(2) = 06 High	PRP 03 – Preventive maintenance PRP 05 – Storing, handling & transportation of material PRP 07 – Pest control PRP 09 – Supply of air water & energy	Time & temperature control Time = 20 min Temp = 124 ±2°C
	P	Extraneous matters (Ash, wood, metal, plastic & dust)	S(2) * P(3) = 06 High	Tea shall be free of Iron according to SLS 1315 part 2	Iron detectors are attached to conveyers at oven output

Ingredient/ PCM/ Processing Step	Potential Hazard (B, C, P)		Is the identified hazard of sufficient probability and severity to warrant control in HACCP plan?	Justification of Decision (If no, identify the prerequisite program that controls the hazard)	Control Measures (Significant hazards only)
Sifting, winnowing & grading 15 – 25	P	Extraneous matters (Ash, wood, metal, plastic & dust) Increase in moisture	S(2) * P(3) = 06 High S(2) * P(2) = 04 Medium	Tea shall be free of Iron according to SLS 1315 part 2 Less than 7% moisture in the final product (operational limit)	Iron detectors are attached to the colour sorter & winnower Moisture measurement of the final product before packing
Packing & storing 26	B	Invasion of *Salmonella,* *E. Coli* & *Staphylococcus aureus*	*Salmonella* Risk S(3) * P(2) = 06 High *E. Coli* S(3) * P(1) = 03 Low *Staphylococcus aureus* S(3) * P(3) = 09 High	PRP 03 – Preventive maintenance PRP 06 – Personal Hygiene PRP 10 – Work environment PRP 11 – Cleaning & sanitation of plant machinery, equipment & utensils	Moisture measurement of the final product before packing
	P	Extraneous matters (dust, packing materials) pellet formation or granulation	S(1) * P(3) = 03 Low S(1) * P(1) = 03 Low	Use of food-grade packing materials PRP 05 – Storing, handling & transportation of material	
Transportation 27	P	Extraneous matters Increase of moisture	S(1) * P(1) = 01 Low S(1) * P(3) = 03 Low	PRP 03 – Preventive maintenance PRP 05 – Storing, handling & transportation of material	

Note:
P = Prevalence, S = Severity
S*P Possible score combinations:

Low Risk	**- 1*1, 1*2, 1*3, 3*1, 2*1**
Medium Risk	**- 2*2**
High Risk	**- 3*2, 3*3, 2*3**

Selection of control measures are based on following: Risk Level = 1 – 3 Low Risk – Establish Control Measures where Appropriate (Consider Synergetic Effects)
: Risk Level = 4 Medium Risk – Establish Control Measures & Modify PRPs if Necessary
: Risk Level = 6 & 9 High Risk – Further Categorization in to CCP/ OPRP (Use Decision Tree Sequence)

HAZARD CONTROL PLAN – CCP/OPRP DECISION TABLE

Process Step	Significant hazard identified in the Hazard Analysis	Q1. Does the process/non-process step have a significant food safety hazard and/or probability of failure of its functioning? If no. This is a PRP. If yes. Move to Q2.	Q2. Does this step specifically designed and capable of preventing or reduce the likely occurrence of hazard(s) to an acceptable level? If no. Move to Q3. If yes. This is a CCP/OPRP. Move to Q5.	Q3. Could contamination or failure of identified hazard(s) occur in excess of acceptable levels, or could this increase to unacceptable levels? If no. Not a CCP/OPRP. If yes. Move to next Q4.	Q4. Will a subsequent step prevent or reduce the likely occurrence of hazard(s) to an acceptable level? If no. Move to Q5. If yes. Not a CCP/OPRP.	Q5. Does the measurable critical/ control measures or observable action criteria available? If no. Modify SPP[1] & move back to Q2. If yes. Move to Q6.	Q6. Does monitoring is feasible with measurable critical/ Control limits or observable action criteria? If no. Modify SPP[1] & move back to Q2. If yes. Move to Q7.	Q7. Does it a process step or a non-process step? If Process. Move to Q8. If Non-Process step. Move to Q10.	Q8. Does it a single control measure or part of a combination of control measures (CCM)/action criteria (AC)? If single. Move to Q9. If CCM/AC. Move to Q10.	Q9. Does it enable timely correctio ns at a failure? If yes. This is a CCP. If no. Modify SPP[1] & move back to Q2.	Q10. Does it enable control of process or product at a failure? If yes. This is an OPRP. If no. Modify SPP[1] & move back to Q2.	CCP ID #
Fermentation **13**	B - Growth & proliferation of *Salmonella, E. Coli Staphylococcus aureus & Aspergillus*	Yes	Yes	-	-	Yes	Yes	Process	CCM	-	Yes	OPRP-1B*
Drying/Firing **14**	B - survival of *Salmonella, E. Coli Staphylococcus aureus Aspergillus*	Yes	Yes	-	-	Yes	Yes	Process	Single	Yes	-	CCP-2B
Sifting, winnowing & grading 15 - 25	P - Extraneous matters (Iron) P - Increase in moisture	Yes Yes	Yes No	- No	- Yes	Yes	Yes	Yes	CCM	-	Yes	OPRP-1P**
Packing & Storing 26	P - Increase in moisture	Yes	Yes	-	-	Yes	Yes	Yes	AC		No	OPRP-2P***

*Fermentation is a continues from rolling and time, temperature and RH are controlled and it is not practical to measure microbial growth or monitor them, In addition, risk is fairly lower than a CCP under properly monitored conditions and the risk is mitigated in next processing step (Drying/Firing – CCP).

** Iron removal is carried out in different steps and the risk is controlled by number of magnets where exact CCP can't predict since all control points are important. Nonetheless, if there is a failure, the irons can be easily removed by passing through a next magnet to correct the issue and cleaning the magnet will eliminate the failure.

*** Moisture of the final product is continually increases until packing which can't be controlled at one point, and the risk is lower than a CCP till it reaches to a certain level (%). If it reaches higher moisture levels, re-drying of product will easily reduce the moisture and eliminate the potential contaminations.

SUMMARY OF POSSIBLE FOODBORNE ILLNESSES CAUSING MICROBES RELATED TO THE TEA INDUSTRY

Cause of Illness	Source	Hazard	Control / Precaution
Aspergillus flavus Aspergillus parasiticus	Present in most raw, dried, and processed foods, especially in plant materials, nuts, dried fruit, and stored cereal grains	They produce aflatoxins. Toxins are heat stable and concentrating, thus stay in the food along the food chain, unaffected by heat treatments. Aflatoxins are genotoxic it affects the genetic material. Genotoxins have a direct dose-response relationship and no threshold dose to exceed.	There is no tolerable daily intake (TDI) for aflatoxins, which are to be kept at a level as low as possible. Though, maximum levels are set in the European Union, for example, 0.05 µg/kg of milk for aflatoxin M1.
Bacillus cereus	Present in most raw, dried, and processed foods, especially spices, cereals, and starchy foods	Spores survive cooking and grow well in foods such as cooked rice dishes. Growth of the bacteria occurs best at room temperatures which may produce two toxins that survive further cooking. One causes diarrhoea, another vomiting	Cool pre-cooked foods quickly in shallow dishes (< 10 cm) and Store below 4°C. Alternatively, hold food above 60 degrees C. Do not store pre-cooked rice, etc. at room temperature
Clostridium botulinum	Spore forming bacterium found almost everywhere, especially in soil and agricultural products	Spores survive and grow in inadequately cooked canned and preserved foods such as meats and vegetables, Usually fatal unless quickly treated. Recovery may take years.	Approved thermal processes for canning and preserving food. Acidification of preserved food May sometimes grow in deep containers in the absence of oxygen. Never store stews etc., in deep pots.
Escherichia coli Enterotoxigenic Enteropathogenic Enteroinvasive Enterohaemorrhagic	Enteric. The types that cause illness are usually of human origin; Certain types commonly in cattle, beef, and raw milk	Transmitted through poor personal hygiene and inappropriate handling of foods Some types cause serious illness and death	Scrupulous cleanliness, especially after using the toilet. Thorough cooking. Control of cross-contamination
Salmonella (non-typhi)	Wide range of animals and foods of animal origin - esp. poultry and meats; also infected persons/ carriers/food handlers	As for *Campylobacter* and *E. coli.* May rarely be fatal	Thorough cooking, rapid cooling & control of hygiene, and cross-contamination
Salmonella typhi and *paratyphi*	Humans (enteric)	as above May be fatal	as above Exclude carriers from food preparation
Shigella	Humans (enteric)	Transmitted through poor personal hygiene and inappropriate handling of foods	Scrupulous cleanliness, especially after using the toilet. Thorough cooking.
Staphylococcus aureus	Humans; from the skin, hair, lesions (cuts and sores), nose & throat	Growth in food causes toxin in food which is not destroyed on reheating	Personal hygiene, thorough cooking, and quick chilling
Viral Gastroenteritis (e.g., Norwalk)	Humans	Handling (may be airborne - coughs, sneezes, etc.)	Scrupulous hygiene, cooking, and storage

AFLATOXINS – *RISK PROFILE*
INTRODUCTION
Mycotoxins are produced by mold, and there is a range of different compounds originating from different types of mold among the Deuteromycetes. The Mycotoxins are secondary metabolites, toxic in low concentrations in vertebrates. **Aflatoxins** are a group of mycotoxins produced by *Aspergillus flavus* and *Aspergillus parasiticus*. These may grow in plant leaves, nuts, dried fruit and infect stored cereal grains where they produce the aflatoxins. When such toxins are formed, they do not go away. They are heat-stable and thus stay in the food along the food chain, unaffected by heat treatments such as pasteurization. When a cow eats a feed contaminated with aflatoxin B1, the activity of the cow changes the aflatoxin B1 to aflatoxin M1, which ends up in the milk (Johnsson, 2006). **The toxicity** of aflatoxin is mainly due to its carcinogenicity. This is because aflatoxins are genotoxic, meaning it affects the genetic material. Genotoxins have a direct dose-response relationship, so they do not have a threshold dose to exceed before they have an effect. Thus, there is no tolerable daily intake (TDI) for aflatoxins, which are to be kept at a level as low as possible. Though, maximum levels are set in the European Union, for example, 0.05 µg/kg of milk for aflatoxin M1. **Mycotoxins** in the product can only be prevented at the source where the mold infection occurs since the toxins are impossible to remove. In the case of made tea, control measures could be taken at the production and handling. **Except** for the opinions, there are no proper examples available to prove aflatoxin contamination in made tea. Thus, there is a very minimal risk for getting the aflatoxins into the made tea. Since aflatoxins have a cumulative negative effect on humans, they shall be considered as hazards in the HACCP plan of the ISO 22000 FSMS.
DISEASE, SYMPTOMS, AND ONSET
Causes Cancers have a cumulative negative effect on humans. Mycotoxins are secondary metabolites, toxic in low concentrations in vertebrates.
MAIN DISEASE FACTOR
Accumulation of aflatoxin.
SOURCE
Post-heat-processing contamination from plant environment including plant personnel, equipment, floors, walls, drains, condensation from coolers, etc.
TRANSMISSION
Consumption of contaminated, processed ready-to-eat nuts and fruits.
CHARACTERISTICS
Heat stable and thus stay in the food along the food chain, Unaffected by heat treatments as pasteurization.

SALMONELLA SPP - RISK PROFILE

INTRODUCTION

***Salmonella* spp**. is a gram-negative, facultatively anaerobic organism, which does not form spores. Growth occurs at 5 - 47°C, and the organism is heat sensitive. It is a zoonotic organism, which may be found in different animals' guts.

Because *Salmonella* may be found in animals, it has a good chance of contaminating foodstuff of animal origin, like meat, milk, and egg. **For example,** in Sweden and Finland, there is zero tolerance regarding *Salmonella* infection among domestic animals, which lowers the risk of human infection. Good handling and heat treatment of the food is necessary to decrease the risk.

Salmonella may cause either enteritis or systemic infection. The gastrointestinal infection has symptoms as milk fever, vomiting, and diarrhoea lasting for a few days up to more than a week. The infectious dose is in the order of 10^6 cells, but in some cases, it has been much lower than that. The systemic infection is caused by invasive, host-adapted serotypes. The bacteria then spread in the body and causes fever, headache, and diarrhoea. *Salmonella* infection may be fatal.

The tea leaves are plucked and heaped on the floor or in the field and transported to the factory with a high risk of contamination from humans due to handling. The raw material further directly handled by many operators during withering, rolling, fermentation, sifting, and packing, which increases the risk of *Salmonella* infection. Hence, there is a risk for *Salmonella* contamination in the product. Luckily the organism's heat sensitivity makes it possible to eliminate with LTHT treatment. The severity of the disease in combination with the risk makes *Salmonella* a hazard to target in the HACCP plan.

DISEASE, SYMPTOMS, AND ONSET

Causes acute diarrhea, vomiting, abdominal pain, and fever. Occasionally, it may cause bloodstream infections and death. Symptoms occur 6 to 72 hours after eating contaminated food.

MAIN DISEASE FACTOR

Bacterial invasion of the lining of the intestine.

SOURCE

Faecal contamination by meat and poultry and maybe by the swage lines.

TRANSMISSION

Primarily from consumption of raw or uncooked eggs, milk, meat, and poultry. Consumption of contaminated, processed ready-to-eat meat products. It also contains in unpasteurized or contaminated dairy products.

CHARACTERISTICS

Killed by mild heat.
Grows with or without air - however, grows best at human body temperature
Very poorly grows at refrigerated temperatures and does not grow above 130°F.
Acid conditions inhibit the growth at all or in most cases.
It survives well in frozen or dry foods, and bacteria in dry foods are more resistant to heat.

E. COLI - *RISK PROFILE*
INTRODUCTION
Escherichia coli are in the family Enterobacteriaceae, gram-negative, rod-shaped, non-spore forming, and motile or non-motile. They can grow under aerobic and anaerobic conditions where they grow best at 37°C. Therefore it is easy to eradicate by simple boiling or basic sterilization. *E. coli O157:H7* is a well-studied strain of the bacterium *E. coli*, which produces Shiga-like toxins, causing severe illness. *E. coli* is transmitted to humans primarily through the consumption of contaminated foods, faecal contamination of water, and other foods. The infectious dose of *E. Coli* is 10^6 - 10^8 logs of organisms. The tea leaves are plucked and heaped on the floor or in the field and transported to the factory with a high risk of contamination or cross-contamination from humans due to handling. The raw material further directly handled by many operators during withering, rolling, fermentation, sifting, and packing, which increases the risk of *E. coli* infection. Hence, there is a risk for *E. coli* contamination in the product due to poor personal hygiene and improper cleaning of utensils. Luckily the organism's heat sensitivity makes it possible to eliminate with LTHT treatment. The severity of the disease in combination with the risk makes *E. coli* a hazard to target in the HACCP plan.
DISEASE, SYMPTOMS, AND ONSET
Symptoms include abdominal cramps and diarrhea, fever, and vomiting may also occur. The incubation period can range from three to eight days, with a median of three to four days. Most patients recover within ten (10) days, but infection may lead to a life-threatening disease, such as hemolytic uremic syndrome (HUS).
MAIN DISEASE FACTOR
Commonly found in the lower intestine of warm-blooded organisms (endotherms).
SOURCE
E. coli is almost exclusively of faecal origin, and it is transmitted through faecal contamination of foods and water, as well as cross-contamination or by direct human contact during food preparation.
TRANSMISSION
E. coli is transmitted to humans primarily through the consumption of contaminated foods such as faecal contamination of water, as well as cross-contamination during food preparation (contaminated surfaces and utensils). Waterborne transmission has been reported, both from contaminated drinking water and from recreational waters. Person-to-person contact is an important mode of transmission through the oral-faecal route.
CHARACTERISTICS
E Coli can grow in temperatures ranging from 7°C to 50°C. Some EHEC can grow in acidic foods, down to a pH of 4.4, and in foods with a minimum water activity (Aw) of 0.95. It is destroyed by thorough cooking of foods until all parts reach a temperature of 70°C or higher.

STAPHYLOCOCCUS AUREUS – RISK PROFILE
INTRODUCTION
Staphylococcus aureus is a facultative anaerobic organism, which does not form spores. Growth occurs at 7 – 45°C with optimum growth at 37°C, and the organism is heat sensitive. The presence of *Staphylococcus aureus* is of concern in products that are fermented. *Staphylococcus aureus* can multiply to high numbers during fermentation if the product is not rapidly fermented (e.g., the starter culture is not active) and cause a toxin to be produced that can cause illness to consumers. The bacteria are common in the environment and are often found on the skin, nose, mouth, or boils and cuts of people. The product may generally become contaminated with *Staphylococcus aureus* from the raw materials or from human contact. Generally, it takes high numbers and growth of *Staphylococcus aureus* to cause a hazard with a medium dose. The symptoms of *Staphylococcus aureus* food poisoning are nausea, vomiting, stomach cramps, prostration, diarrhoea, and last for 6 to 24 hours. Proper cooking, fermentation, cooling, storage, and personal hygiene of food handlers can prevent growth while minimizing cross-contamination of *Staphylococcus aureus* and, more importantly, the production of their toxins. However, cooking will not destroy toxins once they are formed in food. **The** tea leaves are plucked and heaped on the bare floor or in the field and transported to the factory with a high risk of contamination from humans due to handling. The raw material further directly handled by many operators during withering, rolling, fermentation, sifting, and packing, which increases the risk of *Staphylococcus aureus* contamination. Further, the intermediate product is fermented for a minimum of 2.5 hours and maybe as long as 6 hours in some specific tea grades. The severity of the disease in combination with the risk makes *Staphylococcus aureus* a hazard to target in the HACCP plan.
DISEASE, SYMPTOMS, AND ONSET
Causes vomiting, nausea, abdominal cramps, and diarrhea 1-6 hours after eating food contaminated with a toxin produced by this organism.
MAIN DISEASE FACTOR
Production of a heat-resistant toxin in food.
SOURCE
May be present on green leaves and made tea, but contamination of food is primarily from humans.
TRANSMISSION
Bacteria multiply in food products to high levels and produce a heat-stable toxin.
CHARACTERISTICS
Bacteria are killed by mild heat. Toxins are very heat stable and will withstand thermal processing for prolonged periods. Bacteria grow with or without air at body temperatures. The toxin is not usually produced in acid food. Bacteria are resistant to high salt (up to 15%).

Severity	Damage	Examples
I Catastrophic	Death or lasting damage	Listeriosis (pregnant women, elderly people with an immuno-defect) Botulism Perforation within digestion system by a sharp foreign body
II Critical	Many concerned people and lasting or continuous damage	Due to substances with carcinogenic effect Spongiform encephalophathitis
III Serous	Many concerned people but no lasting or continuous damage	Salmonellosis Poisoning by Staphylococcus toxins Allergic reactions due to antibiotic residues
IV Low	Single case and no lasting damage or minimal concentration	Broken tooth by a foreign body, without effect on the digestion system Minimal pesticide residues
V Ignorable	Hazard to be discovered prior to the consumption or minimal indisposition without a clear link to the consumed product	Voluminous foreign bodies Visible product defects (e.g., moldy product, colour, and/ or odour changes)

Likelihood	
A Very Frequent	1x /Week
B Frequent	1x /Month
C Occasional	1x /Semester
D Rare	1x /Year
E Unlikely	1x /3 Year
F Practically Impossible	1x /10 Year

Risk

Acceptable

Low

Serious

Unacceptable

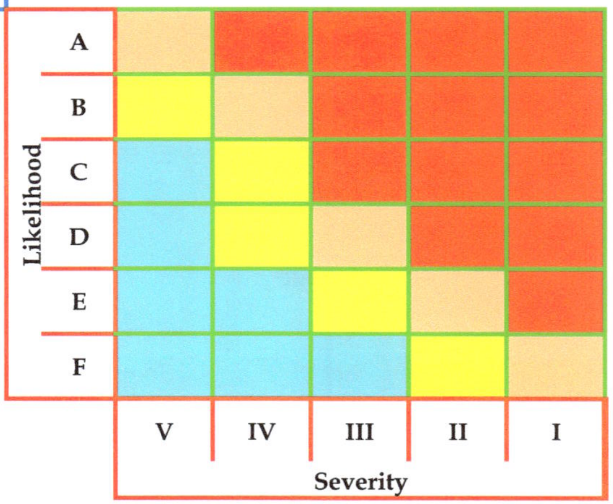

HAZARD CONTOL PLAN – ORTHODOX BLACK TEA

Critical Control Point Plan Summary									
Plant Name	XYZ Tea Factory								
Critical Control Point (CCP)	**Hazard**	**Critical Limit**	**Monitoring**						
			What	**How**	**Frequency**	**Who**	**Corrective Actions**	**CCP verifications**	**Records**
Drying/ Firing	Survival and proliferation of *Salmonella spp.* *S. aureus* *Escherichia coli*	125 ±2° C	Temperature	Reading oven temperature	Every one hour	Operator	Segregate batch for reprocessing or discard (store airtight) Temperature control adjustments	Oven thermometer calibration	FSMS 04/HC/RC/02 Time and Temperature Records
	Over dried tea	20 minutes	No of minutes	Inlet and outlet air quality	Each batch	Operator	Time control Adjustments	Oven retention time test	Calibration records
	Retention of moisture	3% Moisture	Moisture %	Measure Moisture Proper maintenance FSMS 03/WI/FR/03	Each batch	AFO – Firing	Re-dry to the desired moisture %	Moisture analyser calibration Internal audit	Visual and Moisture analysis reports. Internal audit reports

HAZARD CONTROL PLAN – ORTHODOX BLACK TEA

Operational Prerequisite Programme Plan Summary

Plant Name: XYZ Tea Factory

OPRP	Hazard	Operational limit	Monitoring: What	Monitoring: How	Monitoring: Frequency	Monitoring: Who	Corrective Actions	OPRP verifications	Records
Fermentation	Introduction and proliferation of Salmonella spp. Staphylococcus aureus Escherichia coli	2.5 hours 20 ±5° C or 80 ±5° F Hygienic floor	Fermentation Temperature Hygienic condition of the fermentation room and trolleys/ floor	Temperature and R H control Time control FSMS 03/WI/RR/02	Continuous (Every half hour)	Operator AFO – Rolling	Segregate batch in a crate for reprocessing or discard	Thermometer Calibration. Microbiological analysis of intermediate product and floor Internal audit	FSMS 04/HC/RC/01 Time/ temperature/RH record Microbiological analysis reports Calibration records Internal audit reports
Iron removal	Accumulation of Irons through machinery parts of various processing machines and at the field	0% irons at the final product	Use of iron detectors to collect iron at different processing steps	Cleaning Iron detectors after each batch FSMS 03/WI/FR/03 FSMS 03/WI/GR/04	Before start/end of each shift and start of each batch on the go	The operator in relevant machines AFO – Firing/ Sifting / packing	Segregate batch and pass through winnower to recollect iron	Iron detector efficiency calibration Internal audit	FSMS 04/HC/RC/03 Iron collection record Iron detection efficiency reports. Internal audit reports
Packing and Storing	Increase of moisture	5 ±2 % Moisture	Hygienic condition of the sifting room and packing room R H	Measure Moisture Temperature and R H control FSMS 03/WI/PR/05	Each batch	Operator AFO – Firing/ packing	Segregate batch for reprocessing or discard Temperature control	Moisture analyser calibration Internal audit	FSMS 04/HC/RC/04 Moisture analysis reports. Calibration records Internal audit reports

XYZ TEA FACTORY
HAZARD CONTROL PLAN – FERMENTATION RECORD

Black Tea

D. Ref: FSMS 04/HC/RC/01

ISO 22000 – OPRP PLAN
FERMENTATION RECORD

Date	Batch No	Dhool No	Time		Dhool Thickness	Fermentation Time	Dryer Feeding	Visual Inspection		Temperature °F				Responsibility (A.F.O – Rolling)	FSTL's Comments (Verification)
			In	Out				Colour	Oduor	Dry	Wet	Diff	RH%		
		1													
		2													
		3													
		4													
		B/B													
		1													
		2													
		3													
		4													
		B/B													
		1													
		2													
		3													
		4													
		B/B													
		1													
		2													
		3													
		4													
		B/B													
		1													
		2													
		3													
		4													
		B/B													

Prepared : FSTL
Reviewed : C - FSMS
Approved : D/GM

Revision : 00
Date : 01-11-2018
Page : Page 1 of 1

ISO 22000 – CCP PLAN

FIRING RECORD

Date	Batch No	Firing Time		Temperature		Intake Ticket No.	Weight Kg	Moisture %	Responsibility (A.F.O – Firing)	FSTL's Comments (Verification)
		In	Out	Inlet 240 – 255°F	Outlet 110 - 140°F					

ISO 22000 – OPRP PLAN

IRON COLLECTION RECORD

Date	Time	Machine Name	No of Magnets	Weight Collected			Magnet Cleaned	Responsibility	Magnet Inspected (A.F.O – FR/SR)	FSTL's Comments (Verification)
				Mag 1	Mag 2	Mag 3				

Black Tea

XYZ TEA FACTORY
HAZARD CONTROL PLAN – PACKING RECORD

D. Ref: FSMS 04/HC/RC/04

ISO 22000 - OPRP PLAN

PACKING RECORD

Date		Blend Code	Invoice No	Grade	Unit Weight Kg	Quantity Kg	Moisture %	Auctioneer	Responsibility (A.F.O – Packing)	FSTL's Comment (Verification)
Packed	Despatched									

Prepared : FSTL
Reviewed : C - FSMS
Approved : D/GM

Revision : 00
Date : 01-11-2018
Page : Page 1 of 1

WORK INSTRUCTION

1.0 Subject - Withering

2.0 Objectives

Preparation of leaves for rolling through the removal of water and to change the physical state of the leaf.
Control of temperature to minimize the breakdown of chemical compounds inside the leaf.
Extend the minimum time required up to 8 hours before chemical degradation starts inside the leaf.

3.0 Responsibility - Withering Operator

4.0 Instructions

4.1 Clean yourself and wash your hands before entering the production facility and then wear appropriate uniforms, gloves, masks, and caps.
4.2 Clean the utensils, equipment, and withering area within the first 15 minutes.
4.3 Check the hygrometers for appropriate levels of water and add water if necessary.
4.4 Green leave is weighted and sorted by the appointed personnel, but leaf needs to be tested and verified for matured leaves (B - 60 Leaf Grading/Randahlu Verification) as well as the presence of high moisture.
4.5 Check for the presence of impurities, biological or physical hazards in green leaf.
4.6 Remove impurities, none tea leaves, insects, or dead insects if rarely present.
4.7 If the impurities concentration is considerable, reject the relevant amount with the approval of the relevant officer working in the withering area.
4.8 Stack the withering trough loosely up to the top level of the trough without compressing the leaves.
4.9 Fix the name/code tag on the trough relevant to the Supplier/Area/Vehicle from the starting end of the specific green leaf quantity.
4.10 Start the withering fans after completion of loading the withering troughs.
4.11 Control the hygrometer difference around 4°F - 6°F (6°F - 8°F on a raining day) and supply dry heated air.
4.12 Observe withering time. Initial withering will take place at the bottom layer, which has to be check from time to time.
4.13 Stop the fans when the bottom layer is withered completely, and turning and mixing have to be carried out from time to time and re-stack before re-starting the fan.
4.14 Start releasing appropriately withered green leaves to the rolling room after approval of the relevant officer
4.15 The leaves have to be weighed according to the roll size before releasing to the rolling operation and release when the request is made by the rolling section.

4.16 Always release leaves from a single withering trough and start a new trough once the entire trough is emptied properly.

4.17 Record the start and end timings, turning and mixing times, as well as hygrometer readings appropriately on the given recording formats.

4.18 Clean the withering area, withering troughs, and equipment properly (Follow the instructions given below).

4.19 Prevent any cross-contamination from biological, chemical, or physical hazards at all stages of withering.

4.20 Inform the relevant officer or technician if there is any breakdown, technical failures, or mechanical issues found on machinery or equipment.

5.0 Cleaning of Withering Area

5.1 Cleaning Time

7.30 am – 7.45 am
4.45 pm – 5.00 pm

5.2 Required Number of Workers

Daily	- 05
Weekly	- 05
Monthly	- 02

5.3 Equipment to be Cleaned

Staircase, Floor, and Deck
Through Nets
Trough Fan Blades
Trough Fan Nets
Trough Fan Louvers
Cobwebs on Walls and Ceiling
Windows, Nets and walls/Surfaces
Internal Fixtures and Floor of the Trough
Signboards on Troughs and Equipment

Following Items to be leaned Under Supervision of an Officer

Trough Fan Blades
Trough Fan Nets
Trough Fan Louvers

5.4 Utensils Required for Cleaning

Brooms	- 08
Hand Ekel Brooms	- 04
Ekel Brooms	- 10
Cobweb Brush	- 01
Sponges	- 06
Cleaning Cloths	- 20

5.5 Method of Cleaning

5.5.1 Daily

Sweep and clean the Deck, Staircase, and Floor properly.
Sweep and clean inside fittings and the internal floor of the trough using the hand ekel broom while removing all the leaf particles remaining there.
Sweep and clean trough nets using the hand ekel broom properly.

5.5.2 Weekly

Clean trough fan blades with a wet cloth properly.
Clean trough fan louvers with a wet cloth properly.
Clean the windows, nets, and walls of the loft with a brush and broom.
Remove the cobwebs on the ceiling and walls with a cobweb brush.

5.5.3 Monthly

Wash and clean all the nets on the windows of the loft.

6.0 Visual Standards after Cleaning

There shall be no remaining dirt or dust on the loft.
There shall be no remaining leaf particles inside the trough or on the nets.
There shall be no remaining dust or dirt on the trough fan, fan blades, or louvers.
There shall be no remaining cobwebs on ceilings or on the walls.
There shall be no remaining withered leaves in plastic creates.
There shall be a clean and tidy nice environment.

7.0 Records

Withering Area Cleaning Checklist – FSMS 04/PRP/11/RP/01
Hygrometer Temperature record – FSMS 04/PRP/05/RC/01
Withering Record – FSMS 04/PRP/05/RC/02

8.0 Reference

Control of Monitoring and Measuring of Equipment and Methods - FSMS 02/PRP/04
Storing Handling and Transportation of Materials – FSMS 02/PRP/05
Control of Monitoring and Measuring equipment and methods – FSMS 02/PRP/11

WORK INSTRUCTION

1.0 Subject – Rolling, Sifting, and Fermentation

2.0 Objectives

Preparation of withered tea leaves for optimum chemical reactions through the provision of physical conditions, including moisture and air, to obtain intrinsic colours, flavours, and chemical compositions.

Mixing of oxidative compounds with enzymes which are kept separate to catalyse the chemical reactions through cell wall breakage.

Segregation of different parts of the tea shoot for the creation of different tea grades while providing the twisting effects to tea particles.

3.0 **Responsibility** – Roll and Roll Breaker Operators

4.0 Instructions

4.1 Clean yourself and wash your hands and feet before entering the production area and then wear appropriate uniforms, gloves, masks, and caps.

4.2 Clean the utensils, equipment, and rolling room within the first 15 minutes.

4.3.0 Rolling

4.3.1 Clean the rollers, sifters, crates, utensils, and equipment within the first 15 minutes.

4.3.2 Obtain the available quantities for rolling and calculates the number of batches to be rolled and the green leaf releasing schedule based on the withering condition of the given quantities under the supervision of the supervisor/AFO.

4.3.3 Obtain specific instructions on the rolling times, pressure limits, and on, off times and fermentation methods from the section supervisor/AFO.

4.3.4 Test run the machines for accuracy and optimum working conditions; if there is any malfunctioning or breakdown, inform the supervisor.

4.3.5 Request the withered leaf and check the accuracy of the withered level of the leaves.

4.3.6 Check the accuracy of the amount inserted for rolling and maintain at the optimum amounts give per batch (e.g., 300kg)

4.3.7 Collect the spilled tea leaves out of the roller and insert them back to the roller. Push the tea leaves that are lying on the roller table outside the roller back-in for proper and even rolling.

4.3.8 Record the start and end times on a rolling timetable.

4.3.9 Rolling should be carried out only on the given instruction.

4.3.10 Use the given pressures and on-off times accurately as instructed (i.e., 5 ON : 5 Off, 7 ON : 3 OFF)

4.3.11 Release the rolled tea leaves for the roll breaking after rolling in the first roller through releasing the door.

4.4.0 Sifting/Dhool Separation

4.4.1 Check the roll breaker sieves for any damages or impurities before starting the roll breaker. Test run the roll breaker for accuracy and optimum working conditions; if there is any malfunctioning or breakdown, inform the supervisor.

4.4.2 Spread a given a clean coverlet underneath the sieves of roll breaker or spread the dhool crates properly or both.
4.4.3 Start the roll breaker and supply the optimum amount of rolled green leaves while adjusting the hopper mouth for the separation of planned dhools.
4.4.4 Check the sieving for accuracy and optimum flow after the roll breaker is started and clean the mesh to eliminate clogging of mesh sieves which block the pass-through dhools.
4.4.5 Separate the first dhool as instructed above and apply the same method to segregate 2nd, 3rd and 4th dhools as to the given time of rolling and roll breaking through the rest of the rolling program.
4.4.6 The last part of the rolled leaves, which comes out of the roll breaker's top end, is called big bulk, needs to be separately stored.
4.4.7 Record all the relevant data as instructed on the given recording sheets.

4.5 Clean the roller, roller table, and the roll breaker after every batch is realised and follow the clean as you go procedures, and maintain the machine as well as surrounding area always clean and tidy.

4.6.0 Fermentation

4.6.1 Use given equipment (trays/crates/fermentation racks or trolleys /fermentation beds/continuous fermenters as given with specifications and instruction for proper fermentation.
4.6.2 Obtain the crates out of roll breaker after they are filled up 2.5 inches of height and loosely packed; stack them on the racks for 2.5 hours minimum or as instructed.
4.6.3 Stack the big bulk on fermenting beds or crates up to 3 inches for fermentation; always pack them and spread them evenly and loosely for better air diffusion.
4.6.4 Code the dhools according to the given numbering system and keep records on the given recording sheets.
4.6.5 Control the internal humidity through activation of humidifiers and their water injection rate. Keep the temperature difference between wet bulb and dry bulb to a minimum of 2°F.
4.6.6 Follow the clean as you go procedures, and maintain the equipment as well as surrounding area always clean and tidy.
4.6.7 Release the fermented dhools as the firing program planned for the day and check the accuracy of the colour and the time before releasing any batches.

4.7 Clean the machines, equipment, crates, and sieves after finishing the work, and store coverlets, racks, crates, and equipment in given specific places.
4.8 Record all the relevant data as instructed on the given recording sheets.
4.9 Wash and clean the floor, machines, conveyors, sifters, crates, racks, and surrounding area at the end of the working shift.
4.10 Make sure to practice preventive controls to reduce any microbial, chemical, or physical contamination during rolling and fermentation operations.
4.11 If there are any technical or mechanical defects, immediately inform the technician at work and the AFO.
4.12 Manage a maintenance schedule and preventive maintenance programs as planned (PRP 03 – preventive maintenance).

5.0 Cleaning of Rolling Room

5.1 Cleaning Time

1. 7.30 am – 7.45 am
2. 4.45 pm – 5.00 pm

5.2 Required Number of Workers

1.	Daily	- 07
2.	Weekly	- 07
3.	Monthly	- 02

5.3 Equipment to be Cleaned

Rolls
Roll Breakers
Roll Vibrators
Conveyers/conveyer Brush
Coverlets
Floor
Cobwebs on Walls and Ceiling
Windows, Nets, and Walls/Surfaces
Equipment, Racks, Crates
Bulbs and Bulb Covers
Sign Boards

Following Items to be leaned Under Supervision of an Officer
Machine motors and electrical equipment

5.4 Utensils Required for Cleaning

Brooms	- 10
Hand Ekel Brooms	- 10
Ekel Brooms	- 10
Cobweb Brush	- 02
Dust Pans	- 03
Sponges	- 10
Cleaning Cloths	- 20
Wipers	- 02
Watering hose	- 01

5.5 Methods of Cleaning

5.5.1 Daily

Remove all unnecessary utensils or equipment; store them in the assigned places before starting your day.

Use the recommended methods and equipment given for the specific cleaning activities and clean all the machines, equipment, and utensils in the rolling room, including dustbins and garbage cans, which need to be wash and disinfected daily.

Clean the rolling table, rollers, conveyors, and roll breakers brush and wash them properly.

Brush the roll breakers and remove remaining leaf particles and clean them using a wire brush and wash them with potable water. Use a high-pressure airline and water jet for more convenient and uniform cleaning of the equipment and machines, their underneath, inside, and the floor.

Brush, wipe and clean the roll vibrators and wash them properly.

Remove the conveyor brushes and clean them properly after removing the remaining leaf particles.

Clean the crates after properly brushing them to remove sticky leaf particles and wash them properly.

Wash the floor with a suitable detergent and rewash with potable water, and disinfect daily after use.

Clean the floor, entrances; after cleaning the equipment, machine, and utensils and wash and disinfect if necessary.

Clean the control units and signboards using a high-pressure air jet.

5.5.2 Weekly

Clean the windows, nets, and walls of the rolling room with a brush and broom.

Remove the cobwebs on the ceiling and walls with a cobweb brush.

Clean and wipe the signboards with a wet cloth.

All the bulbs and their covers need to be cleaned using a clean wet cloth.

5.5.3 Monthly

Wash and clean all the nets on the windows.

All the fans and other hanging air ducts, service lines need to be cleaned and wipe with a wet cloth.

6.0 Visual Standards after Cleaning

There shall be no remaining leaf particles, dirt, or dust on the equipment or machines.

There shall be no remaining leaf particles on the nets/sieves of the roll breakers, conveyors, or vibrators.

There shall be no remaining dust or dirt on the floor, windows, and storage areas.

The conveyers and their brush are free of remaining leave particles or dust.

There shall be no remaining cobwebs on ceilings or on the walls.

There shall be no remaining tea particles or dust in plastic crates.

There shall be a clean and tidy nice environment.

There shall be a clean and dry floor.

7.0 Records

Rolling Room Cleaning Checklist – FSMS 04/PRP/11/RP/02

Dhool Production and Fermentation Record – FSMS 04/HC/RC/01

WORK INSTRUCTION

1.0 Subject – Drying and Fibre Extraction

2.0 Objectives

Rapid arresting of fermentation after it reaches the optimum levels and reduction of water percentage up to 2.5 – 3.0% for the preservation of made tea with its freshness for a long period while achieving its characteristic flavours and colour.

Removal of fibre materials and dust remaining in the dried tea.

3.0 **Responsibility** – Dryer Operators
– Fibre Mat/3T Operators

4.0 Instructions

4.1 Clean yourself and wash your hands and feet before entering the production area and then wear appropriate uniforms, gloves, masks, and caps.

4.2 Clean the utensils, equipment, and grading area within the first 15 minutes.

4.3.0 Burner

4.3.1 Check the dryer for optimum working conditions before firing it.

4.3.2 Clean the air pipes using a wire brush and remove the ashes in the oven while cleaning them properly.

4.3.3 Use the appropriately split firewood and dried which should be stacked using FIFO. Always use small pieces of split firewood under dry conditions.

4.3.4 Fire the burner after inserting the weight amount of firewood which should be staked to provide proper ventilation for the fire. Renew the firewood stocks according to a time schedule with weighted firewood and keep the fire at optimum conditions as to the heat required by the dryer.

4.3.5 Once the firing is optimum and ashes are forming, break the charcoal from the firewood using a metal rod.

4.3.6 Open the air chamber doors appropriately to provide an enhanced air supply.

4.3.7 Check the temperature through the oven thermometer and adjust the fan speeds and firing to provide optimum drying conditions.

4.4.0 Dryer

4.4.1 Once the fire reaches optimum conditions at a given temperature, turn on the main fan, and adjust the air chamber's leaver to provide optimum airflow.

4.4.2 Obtain the properly fermented available dhools ready for drying and spread them over-drying trays while adjusting the supply direction and start the dryer.

4.4.3 Once the top layer trays are completely covered with dhools, open the air vents 3/4.

4.4.4 Check the internal drying time (20 minutes) continuously and adjust the fan speed dryer speed if necessary and monitor inlet and outlet temperatures as given.

4.4.5 Check the inlet (90°C or above as specified) and outlet (125±2°C) temperatures, including drying time, batch numbers, start and end times while recording them on the given recording sheets.

4.4.6 Monitor and manage the water percentage of the dried leaves at 2.5 – 3.0%.

4.4.7 Follow the clean as you go procedures, and maintain the machine as well as surrounding area always clean and tidy.

4.4.8 Request QA officer to check and verify the relevant temperature, humidity after dryer reaches optimum conditions.

4.4.9 Before production start at every shift, relevant temperature adjustments and airflow adjustments should be carried out by the technician/operator under the supervisor of the officer-in-charge of the drying room, who should check specific temperatures and humidity with the help of the QA officer.

4.4.10 Weigh the dried dhools and calculate the outturn of the production and record them on the given record book.

4.4.11 Store the dried tea in given places with identification tags for the reduction of temperature up to room temperature and then release for the fibre removal.

4.4.12 After completion of the drying process, keep the main fan running until the internal temperature of the dryer reach 60°C.

4.4.13 Clean the magnets attached to the machine frequently as to the given instructions.

4.4.14 Always keep a clean environment around the air chambers, fireplace, and air pipes.

4.4.15 Clean the machine, equipment, and crates after finishing the work and store the crates and equipment in given specific places.

4.5.0 3 T Machine/Fibre-mat

4.5.1 Clean the machine, utensils, and equipment within the first 15 minutes.

4.5.2 Test run the machine for accuracy and optimum working conditions; if there is any malfunctioning or breakdown, inform the supervisor.

4.5.3 Obtain specific instructions on the variety to be cleaned and separation methods from the section supervisor/AFO.

4.5.4 Obtain the given variety of tea to be sifted and the output composition planned for sifting.

4.5.5 Fill the feeding hopper with an optimum amount of given tea variety and start to supply the optimum quantity of dried tea while adjusting the flow rate.

4.5.6 Follow the clean as you go procedures, and maintain the machine as well as surrounding area always clean and tidy.

4.5.7 Store the semi-graded tea in given places according to the verity after coding or release them to the next processing step.

4.5.8 Clean the machine, equipment, and crates after finishing the work and crates and equipment in given specific places.

4.5.9 Record all the relevant data as instructed on the given recording sheets.

4.5.10 Clean the floor and surrounding area at the end of the working shift.

4.6 Make sure to practice preventive controls to reduce any microbial, chemical, or physical contamination during rolling and fermentation operations.

4.7 If there are any technical or mechanical defects, immediately inform the technician at work and the AFO.

4.8 Manage a maintenance schedule and preventive maintenance programs as planned (PRP 03 – preventive maintenance).

5.0 Cleaning of Rolling Room

5.1 Cleaning Time

1. 7.30 am – 7.45 am
2. 4.45 pm – 5.00 pm

5.2 Required Number of Workers

1. Daily - 04
2. Weekly - 04
3. Monthly - 02

5.3 Equipment to be Cleaned

Ovens
Burner
Heating Chamber and Air Pipes
3T Machine
Fibre-mat Machine
Conveyers/conveyer Brush
Firewood Store and Surrounding Area
Floor
Cobwebs on Walls and Ceiling
Windows, Nets, and Walls/Surfaces
Equipment, Racks, Crates
Bulbs and Bulb Covers
Sign Boards

Following Items to be leaned Under Supervision of an Officer

Machine motors and electrical equipment
Burner and Oven
Air Chamber and Air Pipes
Drying Trays

5.4 Utensils Required for Cleaning

Brooms	- 04
Ekel Brooms	- 02
Scoop	- 01
Cobweb Brush	- 02
Dust Pans	- 03
Sponges	- 10
Cleaning Cloths	- 20
Wire brush	- 10
Air Jet	- 01

5.5 Methods of Cleaning

5.5.1 Daily

Remove all unnecessary utensils or equipment; store them in the assigned places before starting your day.

Use the recommended methods and equipment given for the specific cleaning activities and clean all the machines, equipment, and utensils in the firing

room, including dustbins and garbage cans, which need to be wash and disinfected daily.
Clean the 3T/fibre mat using a brush and air wash with an air jet.
Clean the air pipes with a wire brush and clean the burner while removing charcoal and ash remaining in the burner. Spry water over the floor, internal and external surfaces of the heaters and wipe with clean cloths provided.
Switch on the main fan to remove the remaining tea articles and the dust from the drying chamber.
Clean the outlet door, motor, conveyer, and other equipment properly.
Remove and clean fibre and dust attached to the upper part of the drying chamber.
Clean the deck, staircase, and floor properly.
Remove the conveyor brushes and clean them properly after removing the remaining leaf particles.
Clean the crates after use properly, brushing them to remove sticky leaf particles and wash them properly.
Clean the floor, entrances; after cleaning the equipment, machine, and utensils and wash and disinfect if necessary.
Clean the control units and signboards using a high-pressure air jet.

5.5.2 Weekly

Clean, brush and remove carbon in the firing chamber smoke ducts.
Clean the windows, nets, and walls of the firing room with a brush and broom.
Remove the cobwebs on the ceiling and walls with a cobweb brush.
Clean and wipe the signboards with a wet cloth.
Wash the floor with an approved detergent.
All the bulbs and their covers need to be cleaned using a clean wet cloth.

5.5.3 Monthly

Wash and clean all the nets on the windows.
All the fans and other hanging air ducts, service lines need to be cleaned and wipe with a wet cloth.
Clean the drying trays with a wire brush and remove debris in air vents.
Clean the firewood store and rearrange with FIFO.

6.0 Visual Standards after Cleaning

There shall be no remaining leaf particles, dirt, or dust on the equipment or machines.
There shall be no remaining leaf particles on the dryer, drying trays, or inside.
There shall be no debris material/ashes inside the burner, and air chambers are clean and tidy.
There shall be a properly arranged, FIFO applied firewood store.
There shall be no remaining dust or dirt on the floor, windows, and remaining areas.
The conveyers and their brush are free of remaining leave particles or dust.
There shall be no remaining cobwebs on ceilings or on the walls.
There shall be no remaining tea particles or dust in plastic crates.
There shall be a clean and tidy nice environment.
There shall be a clean and dry floor.

7.0 Records

Firing Room Cleaning Checklist – FSMS 04/PRP/11/RP/03
Drying Time Table – FSMS/04/HC/RC/02
Iron Collection Record – FSMS 04/HC/RC/03

WORK INSTRUCTION

1.0 Subject - Grading

2.0 Objectives

Extraction of different tea grades from the dried tea based on different colours, shapes, and weights according to standard orthodox black tea grades and customer preferences and market requirements.

Cleaning of unnecessary particles in dried tea and reductions and breaking of large particles to match market preferences.

3.0 **Responsibility** - Middleton/Michi/Chota/Winnower Operators
- Colour/separator/Cutter Operators

4.0 Instructions

4.1 Clean yourself and wash your hands and feet before entering the production area and then wear appropriate uniforms, gloves, masks, and caps.

4.2 Clean the utensils, equipment, and grading area within the first 15 minutes.

4.3.0 Middleton Sifter

4.3.1 Clean the sifter, utensils, and equipment within the first 15 minutes.

4.3.2 Test run the machine for accuracy and optimum working conditions; if there is any malfunctioning or breakdown, inform the supervisor.

4.3.3 Obtain specific instructions on the variety to be sifted and separation methods from the section supervisor/AFO.

4.3.4 Obtain the given variety of tea to be sifted and the specific sieve given according to the output composition planned for sifting.

4.3.5 Clean the mesh properly and fix it to the sifter without leaving any gaps or spaces around it.

4.3.6 Fill the feeding hopper with an optimum amount of given tea variety and adjust the hopper mouth and start the Middleton sifter.

4.3.7 Check the sieving for accuracy and optimum flow.

4.3.8 Use the scrappers to clean the sieve while it is operating, but don't bang the mesh to increase sieving and cleaning.

4.3.9 Follow the clean as you go procedures, and maintain the machine as well as surrounding area always clean and tidy.

4.3.10 Store the semi-graded tea in given places according to the verity after coding or release them to the next processing step.

4.3.11 Clean the machine, equipment, crates, and the sieves after finishing the work and store sieves, crates, and equipment is given specific places.

4.3.12 Record all the relevant data as instructed on the given recording sheets.

4.3.13 Clean the floor and surrounding area at the end of the working shift.

4.4.0 Michi Sifter

4.4.1 Clean the sifter, utensils, and equipment within the first 15 minutes.

4.4.2 Test run the machine for accuracy and optimum working conditions; if there is any malfunctioning or breakdown, inform the supervisor.

4.4.3 Obtain specific instructions on the variety to be sifted and separation methods from the section supervisor/AFO.

4.4.4 Obtain the given variety of tea to be sifted and the specific sieve given according to the output composition planned for sifting.

4.4.5 Clean the mesh properly and fix it to the sifter without leaving any gaps or spaces around it.

4.4.6 Fill the feeding hopper with the optimum amount of given tea variety and adjust the hopper mouth and start the Michi sifter.

4.4.7 Check the sieving for accuracy and optimum flow.

4.4.8 Use the scrappers to clean the sieve while it is operating, but don't bang the mesh to increase sieving and cleaning.

4.4.9 Follow the clean as you go procedures, and maintain the machine as well as surrounding area always clean and tidy.

4.4.10 Store the semi-graded tea in given places according to the verity after coding or release them to the next processing step.

4.4.11 Clean the machine, equipment, crates, and the sieves after finishing the work and store sieves, crates, and equipment is given specific places.

4.4.12 Record all the relevant data as instructed on the given recording sheets.

4.4.13 Clean the floor and surrounding area at the end of the working shift.

4.5.0 Chota Sifter

4.5.1 Clean the sifter, utensils, and equipment within the first 15 minutes.

4.5.2 Test run the machine for accuracy and optimum working conditions; if there is any malfunctioning or breakdown, inform the supervisor.

4.5.3 Obtain specific instructions on the variety to be sifted and separation methods from the section supervisor/AFO.

4.5.4 Obtain the given variety of tea to be sifted and the output composition planned for sifting.

4.5.5 Fill the feeding hopper with an optimum amount of given tea variety and start the bucket conveyer to supply the optimum quantity of semi-graded tea while adjusting the hopper mouth and start the Chota sifter.

4.5.6 Check the sieving for accuracy and optimum flow.

4.5.7 Follow the clean as you go procedures, and maintain the machine as well as surrounding area always clean and tidy.

4.5.8 Store the semi-graded tea in given places according to the verity after coding or release them to the next processing step.

4.5.9 Clean the machine, equipment, crates, and the sieves after finishing the work and crates and equipment in given specific places.

4.5.10 Record all the relevant data as instructed on the given recording sheets.

4.5.11 Clean the floor and surrounding area at the end of the working shift.

4.6.0 Winnower

4.6.1 Clean the winnower creates and magnets within the first 15 minutes.

4.6.2 Test run the machine for accuracy and optimum working conditions; if there is any malfunctioning or breakdown, inform the supervisor.

4.6.3 Obtain specific instructions on the variety to be winnowed and separation methods from the section supervisor/AFO.

4.6.4 Obtain the given semi-graded variety of tea to be winnowed and the specific weights to be achieved according to the output composition planned for winnowing.

4.6.5 Adjust the vibrators accurately for even material supply for winnowing.

4.6.6 Fill the feeding hopper with an optimum amount of given tea variety while opening and adjusting the louvers and start the winnower.

4.6.7 Compare the leave sizes and weights of the first and second crates and optimize louver openings.

4.6.8 Follow the clean as you go procedures, and maintain the machine as well as surrounding area always clean and tidy.

4.6.9 Store the winnowed semi-graded tea as to the weights in given places according to the verity after coding or release them to the next processing step.

4.6.10 Clean the magnets attached to the machine frequently as to the given instructions.

4.6.11 If winnowing is not optimum, check the dust bags and clean them properly.

4.6.12 Clean the machine, equipment, and crates after finishing the work and store crates and equipment in given specific places.

4.6.13 Record all the relevant data as instructed on the given recording sheets.

4.6.14 Clean the floor and surrounding area at the end of the working shift.

4.7.0 Colour Separator

4.7.1 Clean the colour separator, including conveyors, creates and magnets within the first 15 minutes.

4.7.2 Test run the machine for accuracy and optimum working conditions; if there is any malfunctioning or breakdown, inform the supervisor.

4.7.3 Obtain specific instructions on the variety to be cleaned and separation methods from the section supervisor/AFO.

4.7.4 Obtain the given semi-graded variety of tea to be cleaned for impurities and the specific grades/cleaning to be achieved according to the output composition planned for visual uniformity and cleanliness.

4.7.5 Adjust the conveyer accurately for even material supply for cleaning.

4.7.6 Fill the feeding hopper with the optimum amount of given tea variety while opening and adjusting the air supply and start the colour separator.

4.7.7 Compare the required leave composition, cleanliness, colours, sizes, and weights while adjusting the separation, supply, and speed of the machine.

4.7.8 Store the colour segregated, cleaned, and semi-graded tea as to the grades in given places according to the verity after coding or release them to the next processing step.

4.7.9 Clean the magnets attached to the machine frequently as to the given instructions.

4.7.10 Follow the clean as you go procedures, and maintain the machine as well as surrounding area always clean and tidy.

4.7.11 If cleaning is not optimum, use a high-pressure air get and clean the machine properly.
4.7.12 Clean the machine, equipment, and crates after finishing the work and store the crates and equipment in given specific places.
4.7.13 Record all the relevant data as instructed on the given recording sheets.
4.7.14 Clean the floor and surrounding area at the end of the working shift.

4.8.0 Tea Cutter

4.8.1 Clean the tea cutter and creates within the first 15 minutes.
4.8.2 Test run the machine for accuracy and optimum working conditions; if there is any malfunctioning or breakdown, inform the supervisor.
4.8.3 Obtain specific instructions on the variety to be sized and specific instructions from the section supervisor/AFO.
4.8.4 Obtain the given semi-graded variety of tea to be sized and the specific grades to be achieved according to the output composition planned.
4.8.5 Fill the feeding hopper with the optimum amount of given tea variety while opening and adjusting the supply flow and start the cutter.
4.8.6 Store the sized/cut and semi-graded tea as to the grades in given places according to the verity after coding or release them to the next processing step.
4.8.7 Follow the clean as you go procedures, and maintain the machine as well as surrounding area always clean and tidy.
4.8.8 Record all the relevant data as instructed on the given recording sheets.
4.8.9 Clean the floor and surrounding area at the end of the working shift.

5.0 Cleaning of Grading Area

5.1 Cleaning Time

1. 7.30 am – 7.45 am
2. 4.45 pm – 5.00 pm

5.2 Required Number of Workers

1. Daily – 20
2. Weekly – 10
3. Monthly – 10

5.3 Equipment to be Cleaned

Middleton
Michi/Chota
Winnower
Colour Separator
Conveyors/Conveyor Brush
Floor
Cobwebs on Walls and Ceiling
Windows, Nets, and Walls/Surfaces
Equipment Rack, Weighing Scales,
Bulbs and Bulb Covers
Sign Boards

Following Items to be leaned Under Supervision of an Officer
Machine motors and electrical equipment

5.4 Utensils Required for Cleaning

Brooms	– 15
Hand Ekel Brooms	– 25
Ekel Brooms	– 10
Cobweb Brush	– 03
Dust Pans	– 05
Sponges	– 10
Cleaning Cloths	– 20
Hand Spry	– 01

5.5 Methods of Cleaning

5.5.1 Daily

Remove all unnecessary utensils or equipment; store them in the assigned places before starting your day.
Use the recommended methods and equipment given for the specific cleaning activities and clean all the machines, equipment, and utensils in the grading room, including dustbins and garbage cans, which need to be wash and disinfected daily.
Remove the sieves of the Middleton, Michi, and Chota sifters and clean them using a wire brush to remove any remaining made tea particles or dust and store them or reassemble them properly. Use a high-pressure airline for more convenient and uniform cleaning of the equipment and machines, their underneath, inside, and the floor.

Clean the dust bags of the Winnower, clean the machine, crates, and conveyor belts using a pressurized air jet.
Clean the Colour Separator using a brush and high-pressure air jet, including conveyors, crates, feeding hopper, and other parts.
Remove all the remaining tea leaves or dust and clean the cutter, including the hopper bin, using a high-pressure air jet.
Clean the magnets attached to the machines using given instructions while collecting and weighing attached metal amounts and recording the readings of the day.
Clean the floor, entrances; after cleaning the equipment, machine, and utensils and wash or disinfect if necessary.
Clean the conveyers using a clean piece of cloth and a brush
Clean the control units and signboards using a high-pressure air jet.

5.5.2 Weekly

Clean the windows, nets, and walls of the grading room with a brush and broom.

Remove the cobwebs on the ceiling and walls with a cobweb brush.
Clean all empty crates using an air jet; if they need to be washed, then use a separate washing and drying place with appropriate resources.
Clean and wipe the signboards with a wet cloth.
All the bulbs and their covers need to be cleaned using a clean wet cloth.

5.5.3 Monthly

Wash and clean all the nets on the windows.
All the fans and other hanging air ducts, service lines need to be cleaned and wipe with a wet cloth.

6.0 Visual Standards after Cleaning

There shall be no remaining dirt or dust on the equipment or machines.
There shall be no remaining leaf particles on the nets/sieves, and they shall be stored according to the given instructions.
There shall be no remaining dust or dirt on the floor, windows, and storage areas.
The conveyers and their brush are free of remaining leave particles or dust.
There shall be no remaining cobwebs on ceilings or on the walls.
There shall be no remaining tea particles or dust in plastic crates.
The made tea or semi-graded tea shall be stored and kept on FIFO with coding for traceability.
There shall be a clean and tidy nice environment.

7.0 Records

7.1 Grading Area Cleaning Checklist – FSMS 04/PRP/11/RP/04
7.2 Iron Collection Record FSMS/04/HC/01/03
7.3 Packing Record – FSMS 04/HC/01/RC/04

WORK INSTRUCTION

1.0 Subject – Blending and Packing

2.0 Objectives

Blending of segregated different small quantities of made tea varieties to create even and unique composition according to market preferences while increasing the quantity.

Packaging of products according to auction requirements and traceability.

3.0 **Responsibility** – Blending & Packing Operator

4.0 Instructions

4.1 Clean yourself and wash your hands and feet before entering the blending area and then wear appropriate uniforms, gloves, mask, and cap.

4.2 Clean the utensils, equipment, and blending and packing area within the first 15 minutes.

4.3 Sterilize the blending floor and mixing shovels with IPA (Isopropyl Alcohol) using a sprayer.

4.4 Draw samples from each made tea varieties intended for blending and request a sample test for moisture content (%) and approval before blending from the supervisor/AFO, who is responsible for the equipment operation. Keep the relevant moisture records.

4.5 Weigh the given quantities of same made tea verity (i.e., FBOPF) extracted from 1st, 2nd, 3rd dhools and the big bulk which are intended for blending and packing.

4.6 Wear specific boots or socks given for blending operation.

4.7 Each tea crate needs to be held by two workers from two sides while holding above the shoulders and let it free fall while moving in a circle to create a ridge with empty space in the middle. Make sure all the crates intended for blending are added to the circle with free fall.

4.8 Shovel the ridge of made tea into the middle empty space while creating a heap and check if the blending is enough. If not, further shovelling needs to be carried out to achieve even blending.

4.9 Check the printed paper sacks (variety specific) for accurate printing of the net weight, tare weight, gross weight, grade, bag no, invoice no, selling mark, and registration number (Nett, Tare, Gross, Grade, Bag No, INV No, Selling Mark & MF).

4.10 Fill up the specifically printed and verified sacks according to the printed weight and weigh them accurately.

4.11 Seal the bag mouth with food-grade gums given and store them in the given place after proper fastening of the mouth.

4.12 Check the lorry for the cleanliness and hygienic conditions before loading

4.13 Stack the lorry according to the loading plan given by AFO.

4.14 Update all the relevant records related to blending and packing.

5.0 Cleaning of Blending & Packing Area

5.1 Cleaning Time

7.30 am – 7.45 am
4.45 pm – 5.00 pm
Just before blending starts

5.2 Required Number of Workers

Daily	– 06
Weekly	– 01
Monthly	– 02

5.3 Equipment to be Cleaned

Floor
Cobwebs on Walls and Ceiling
Windows, Nets, and Walls/Surfaces
Equipment Rack, Weighing Scales, Sealers
Sign Boards

Following Items to be leaned Under Supervision of an Officer

Sterilization of Floor
Weighing Scales

5.4 Utensils Required for Cleaning

Brooms	– 03
Cobweb Brush	– 01
Sponges	– 06
Cleaning Cloths	– 20
Hand Spry	– 01

5.5 Method of Cleaning

5.5.1 Daily

Sweep and clean the floor properly.

5.5.2 At Each Blending

Sweep and clean the floor properly.
Use the IPA with accurate concentration.
Spry the IPA using a spryer and wait for few minutes to sterilize the floor and evaporate the residues.

5.5.3 Weekly

Wash the floor properly.
Clean weighing scales with a wet cloth properly.
Clean equipment rack and signboard with a wet cloth properly.
Clean the windows, nets, and walls with a brush and broom.
Remove the cobwebs on the ceiling and walls with a cobweb brush.

5.5.4 Monthly
Wash and clean all the nets on the windows.

6.0 Visual Standards after Cleaning

There shall be no remaining dirt or dust on the floor.
There shall be no remaining made tea/dust inside the equipment racks or weighing scales
There shall be no remaining cobwebs on ceilings or on the walls.
There shall be no remaining made tea particles in plastic creates.
There shall be a clean and tidy nice environment.

7.0 Records

Grading Area Cleaning Checklist – FSMS 04/PRP/11/RP/04
Packing Record – FSMS 04/HC/01/RC/04

1.0 Subject – Control of Health

2.0 Objectives:
To assist in the prevention of cross-contamination, food handlers

3.0 Responsibility – Quality Officer

4.0 Activities:

4.1 Establish a Personnel hygiene inspection record book or control logbook; maintain the record of inspections on a weekly or biweekly basis.

4.2 Food handlers shall maintain a high degree of personal cleanliness and wear suitable protective clothing, head covering, and footwear as per the protective garment guide for each section.

4.3 Each food handler shall be provided with a clean uniform daily, and the used/ dirty uniforms shall be put in dirty clothes bins; that shall be rewashed by themselves after the work. All food handlers shall wash their hands whenever they engage in food processing

4.4 An Assistant Factory Officer shall check the employees while they enter the production area at the beginning of the shift. The AFO will be looking into the following criteria at the entrance. In addition, he will consider suitable requirements from the list as to the area of work.

4.5 He will maintain records for all the employees engaged in food handling in critical areas.

1. Biweekly Health Check Record Book

Health Check Record Book					
Plant Name		Department			
D. Ref	FSMS 04/PRP/06/RC/01	Responsible Officer			
	Aspect	Emp 1	Emp 2	Emp 3	Emp 4
1	Trimmed hair				
2	Trimmed nails				
3	Trimmed moustache				
4	Compliance with ornament policy				
5	Wearing proper uniform				
	Hair cap				
	Mask				
	Apron				
	Slippers				
6	Hand washing				
7	Reporting of illnesses				
8	Visually observed illnesses				
9	Wounds and prevention steps				
10	Any inappropriate behaviours (smoking, chewing)				

2. Protective Garment Register

Protective Garment Supply Register							
Plant Name			Department				
D. Ref	FSMS 04/PRP/06/RG/01		Responsible Officer				
Employee Name/ EPF No:	Uniform	Apron/ Over Coat	Cap	Mask	Gloves	Boots	Goggles

3. Illness and Injuries Register

Illness and Injuries Register						
Plant Name			Department			
D. Ref	FSMS 04/PRP/06/RG/02		Responsible Officer			
Date	Name/ EPF No	Reason(illness or injury)	Informed to	Work point	New work point	Dates removed from working point

1.0 Objective

The objectives of the test procedure to validate and measure the critical control point, to ensure its continuing suitability, adequacy, and effectiveness, assessing opportunities for improvements and changes of the critical control points, operational prerequisite programmes, and prerequisite programs before and after implementing the system.

2.0 Scope

The scope covers all the activities related to the validation of the food safety management system of XYZ Tea Factory.

3.0 Responsibility

Quality Officer	- Initiation and testing of relevant CCPs and OPRPs
Asst. Factory Officers	- Assisting the QO invalidation of control measures
	- Conducting performance verifications as stipulated
	- Notifying QO on validations of CCPs and OPRPs/PRPs
Technician	- Follow the guidelines for mechanical equipment tests
Consultant - FSMS	- Conducting Validation
	- Compliance with FSMS

4.0 Activities

4.1.1 The plant shall conduct initial validation with the help of an external expert/consultant using recognized methods for validation. If any newly designed or modified techniques are used, all such validation protocols shall be tested for accuracy with relevant microbiological, chemical, and physical requirements based on the industry stipulated requirements and norms.

4.1.2 Once the validation methods are proven against reputed industry standards, they can be used to validate the CCPs, OPRPs, PRPs, or any other relevant, measurable, or observational criteria specified for the FSMS. After initial validation is completed, the plant needs to implement the monitoring and verification protocols. Sampling and monitoring levels should be adjusted to a statistically appropriate and representative benchmark.

4.1.3 The monitoring shall be designed to provide evidence that the FSMS is operating as intended, which is carried out in a real-time mode. The monitoring shall be designed to determine if the processes are in control or if a critical limit is exceeded and to initiate corrections and corrective actions immediately.

4.1.4 The temperature of the dryer shall be plotted on a statistical chart using a sensor that will directly send the data to a computer, and statistical process shall determine the violation of critical control limits, and the system shall immediately send alerts to the relevant responsible people while initiating the corrections. The chart shall point out the positive and negative trends without human factor involvement.

4.2 Prior to the implementation of the food safety management system, each operational PRP and CCP shall be validated to determine if it is capable of achieving the intended control of the identified hazards.

4.2.1 According to the obtained results, the control measures or combinations of control measures shall be demonstrated that they are capable of ensuring the defined acceptable levels of the end products.

4.2.2 The food safety management system shall demonstrate that it is designed to produce the desired controls, is operated as designed, and is updated as new information is provided.

4.2.3 The food safety management system shall be developed using sound scientific principles. The necessary information for the system design can usually be obtained from colleges and universities, government agencies or research branches, trade or industry associations, consultants, or other parties that have expertise in the food manufacturing process and product.

4.2.4 Once the control measures are designed on paper; they shall be validated. The validation process provides assurance that the control measure or the combination of control measures will deliver end products that are safe (i.e., within the acceptable levels required for each identified hazard).

4.2.5 Validation of the system shall usually include activities such as:

a. Reference to validations carried out by others or historical knowledge;
b. Experimental trials to mimic process conditions;
c. Collection of biological, chemical, and physical hazard data during normal operating conditions;
d. Statistically designed surveys;
e. Generally accepted industrial practices;
f. Mathematical modelling;
g. Microbiological or analytical testing
h. Scaling up laboratory-based experimental trials in a pilot plant;
i. Intermediate and/or finished product sampling and testing;
j. Use of statistical sampling plans;
k. Validated testing methodology;

4.2.6 In addition, if additional control measures, new technology or equipment, changes in the control measures, identification of new or emerging hazards or their frequency of occurrence, or unexplained failures of the system occur, re-validation of the system shall be necessary.

4.2.7 M/s XYZ Tea factory used to find data available from formal bodies while conducting internal assessments to find the suitability of the current practices. In addition, a checklist will be implemented to find the adequacy of total FSMS covering:

a. Product and process;
b. Product process history;
c. CCPs, critical limits, monitoring, corrective action, CCP verification, OPRPs, and record-keeping procedures. Review current CCP documentation;
d. PRPs of any change or if any modification is required and their impact on food safety.

4.2.8 Validations may be conducted by external parties.

4.3 The main CCP is drying while fermentation, iron collection, and moisture testing become OPRPs in orthodox black tea manufacturing due to their criticality and control measures available.

4.4 **Dryer Testing**

The dryer testing shall be conducted for three parameters which are temperature, residence time, and moisture content.

4.4.1 Temperature of the inlet temperature gauge of the dryer and back end temperature gauge shall be tested with an externally calibrated thermometer while the dryer is operating.

4.4.2 The temperature of both gauges shall be tallied to the calibrated thermometer temperature.

4.4.3 The temperature of the inlet shall be within 88 – 91^{0}C for the endless chain dryer (ECD), and at the exhaust, it shall be within 49 - 54^{0}C. However, the inlet temperature can be set around 100 – 110^{0}C as to the requirements. Thus it is very important to verify the lower end of inlet temperature, which shall be set over 91^{0}C with over 20 minutes of residence time to achieve 3% moisture, appropriate black colour, and proper sterilization of the end product (The fluidized bed dryer shall use 127^{0}C as the inlet temperature while setting discharge from drying section around 88 – 93^{0}C).

4.4.4 The residence time shall be tested by inserting a tag to the ongoing fermented dhool and calculating the time using a stopwatch to see the tag out from the exhaust.

4.4.5 Based on the time calculation, speed shall be adjusted to match the requirements of residence time, which shall be 21 minutes for the ECD with an inlet temperature of 88 – 91^{0}C. This can be varied according to the dhool, manufactured end product, and the blackness required.

4.4.6 The moisture percentage of the end product shall be tested after these two parameters are tested to verify dryer efficiency, which shall be carried out using a calibrated moisture meter.

4.4.7 According to the standard test method of moisture meter, get the reading of moisture percentage of a sample, which shall be within the 2.5 – 3%.

4.4.8 If the above values are not in the company set ranges for different products, they shall be adjusted according to the requirements before manufacturing is carried out through the dryer.

4.5 Smoke Test

4.5.1 The smoke test shall be carried out to verify the dryer performance and if there is any smoke contamination to the end product through the dryer while drying inside the drying chamber of the dryer.

4.5.2 The smoke test shall be carried out once a month by QO/ AFO – firing with the help of a technician/dryer operator using the following method.

a. Remove ashes and dust particles in the heater tubes as explained in the work instructions.
b. Cover the two openings of the heater with clay.
c. Close the ID fan damper to prevent air from flowing to the chimney.
d. Enter firewood with any kind of high smoke generating leaves to the firing chamber of the dryer.
e. Check for any smoke leeks in the drying chamber.
f. If there are not any fumes or odours related to the fired leaves outside, the heater in good condition.

g. If the smoke is leaking to the chamber, then immediately repair the dryer for all the leeks detected.

4.6 **Fermentation**

4.6.1 The fermentation shall be controlled through time, temperature, and relative humidity. The equipment shall be validated internally or externally based on standard methods, time is a general measure, but the equipment used for the specific operation shall be verified with an accurate source.

4.6.2 The hygrometers equipped in the fermentation room shall be verified with an externally calibrated hygrometer by comparing the results and adjusting the error or externally through an accredited laboratory.

4.6.3 Microbial tests for internal air shall be conducted to find the presence of air-born pathogens or spores in the fermentation room.

4.6.4 Dhool samples and operative's hands and contact equipment shall be analysed to find out microbial contamination, growth, and potential for pathogenic bacteria and fungi inside the room.

4.6.5 The fermentation tables, racks, trolley tray surfaces, or cement floors that are used for spreading shall be verified with microbiological swabs after cleaning to verify the effectiveness of cleaning.

4.6.6 The fermentation time is very critical where generation of toxic compounds by bacteria shall be controlled through time and minimization of cross-contamination can be achieved through control of workplace hygiene and personnel hygiene; otherwise, fermentation also can be considered as OPRP.

4.7 **Iron Removal**

4.7.1 The contamination of iron (metal filings) is one of the major physical contamination points where the collection of iron is critical to consumer safety which happens throughout the manufacturing process due to the different machines used and their metal parts used for various processing requirements.

4.7.2 The iron is collected through metal collectors, which are basically made out of magnets. The magnet capacity shall be validated and verified before its utilization. Iron removal is not a single process, and the orthodox process has equipped three magnets within the process. First magnet at dryer mouth, second magnet at winnower, and the third magnet at colour sorter. If there is a situation of all three magnets in a given production have saturated while processing, the product shall be segregated and reprocessed to remove irons in it. Since the XYZ tea factory has two dryers, two winnowers, and three colour separators, all the equipment is equipped with magnets. Thus each magnet shall be verified applying the following method to find out magnet capacity.

4.7.3 Magnets shall be validated through;

The magnets shall be validated first, and then they shall be evaluated for performance. Thus use an electronic weighing scale and a powder funnel as equipment with a known weight of metal filings (e.g., 100grms) and refused tea (e.g., 5kg). The performance checks shall be carried out once every month.

1. Weigh 100g of metal filings and mix it with 5kg of refused tea.
2. Pass the refused tea through calibrating magnet and collect the metals carefully by scrubbing the magnet and weigh the collected filings and verify the amount.
3. If all metal filings have been collected (100grms), increase the amount of metal filings (e.g., 150grms) and follow the procedure again.
4. When the magnet is saturated (e.g., 125grms), carry out the same procedure for three times and get the average.
5. This number shall be considered as the magnet capacity.
6. Each every magnet shall be validated in the same procedure, and it shall be tallied with the iron collection of every magnet if the magnets are observed with a high collection of iron particles.
7. Every moth magnets shall be checked for performance, and if the performance is less than 75% of the capacity, such magnets shall be replaced with a new magnet, complying with similar capacity.

4.8 Calibration of Moisture Meter

4.8.1 The moisture meter shall be calibrated according to the given calibration method with the equipment.

4.8.2 The equipment suppliers provide calibration services with such high-end equipment where it is advised to use their service. Otherwise, use the guidelines provided with the user manual.

4.8.3 If the equipment shows error readings or if suspected for such readings, it shall be calibrated with an accredited external laboratory.

4.8.4 Dried tea shall be checked for moisture at the dryer mouth and after blending before packing to verify the moisture rise and final moisture value of the end product.

4.9 PRP Validation

4.9.1 The PRP's are set for non-critical areas of the manufacturing process, however violation of one PRP may not become a very critical issue to the production, but two or three PRPs may cause adverse health effects to the end-user, where it shall be validated and controlled according to the requirements.

4.9.2 Each PRP shall be assessed for its suitability and applicability to the context of the organization and its place in the food supply chain and asses based on the available reputed data, specific control measures, or against national or international standards.

4.9.3 The PRP shall be modified or verified annually; however, if there is a modification to the production process or specific processing step, the PRP shall be re-evaluated according to the changes.

4.9.4 Maintain a list of PRP's and related documents with their validity and suitability to the current production process.

5.0 References

5.1 ISO 22000:2005 Standard
5.2 Hazard Control Plan
5.3 CCP and OPRP Readings

6.0 Records

6.1	System verification checklist	- FSMS 04/FSP/07/SV/CL/01
6.2	System validation checklist	- FSMS 04/FSP/07/SV/CL/02
6.3	Fermentation Record	- FSMS 04/HC/01/RC/ 01
6.4	Firing Record	- FSMS 04/HC/01/RC/ 02
6.5	Iron Collection Record	- FSMS 04/HC/01/RC/ 03
6.6	Packing Record	- FSMS 04/HC/01/RC/ 04
6.7	Calibration reports	
6.8	Test analysis reports	
6.9	Records of Verification	

1.0 Subject - Control of Emergencies

2.0 Objectives

Creation of awareness among employees on prevention and control of highly potential emergency situations in the tea manufacturing facility while minimizing the damages to the company properties or its employees.

3.0 Responsibility

Trained employees and the executives

4.0 Prevention of Emergency Fires

Control of emergency fires requires a better understanding of critical factors required to occur a fire, type of fires and their major requirements, as well as control mechanisms.

4.1 Requirements

There are three mandatory requirements to fulfill for the occurrence of a fire.

Fuel
Oxygen O_2/ Air
Heat

4.2 Type of Fires

General Fires – solid based fires (wood/paper/cloths/plastic
Oil Fires – liquids or convertible to liquids (petrol/diesel/grease/paints)
Electrical & gas Fires – Fire eruptions due to electricity or gases

4.3 Reasons for Fire Eruptions

Ignorance
Unawareness
Emergency accidents
Natural disasters
Intentionally created fires

4.4 Prevention of Fire Accidents

4.4.1 Follow all the guidelines and instructions for fire accidents and prevention.
4.4.2 Prevent and control smoking inside the factory premises
4.4.3 Store flammable materials away from high temperature generating areas or exposure to such temperatures.
4.4.4 Follow the given instructions while using the equipment without fire protection or safety cover.

4.5 Methods of Fire Protection

Smothering
Cooling
Starvation

4.6 Fire Protection Equipment

The fire protection equipment is used at the primary stage of a fire with trained staff, but critical or extended fires require the support of a fire fighting unit. There are different type of fires which require a specific type of fire protection according to the type of fire. The protection equipment has specific colours according to the type of fire, which makes selection easy; thus, use the appropriate colour code based on the fire when required.

1. H_2O Fire Extinguisher – General fires
2. CO2 Fire Extinguisher – Oil/Electrical/Gas fires
3. Form Fire Extinguisher – Oil/General fires
4. Dry chemical powder fire Extinguisher – All type of fires

4.7 Control of Emergency Fires

4.7.1 Once a fire irrupted, inform everyone by shouting loudly "fire."
4.7.2 Ring the emergency fire switches close by.
4.7.3 Inform the relevant officers and trained firefighters in the factory.
4.7.4 Once the alarms are active, evacuate all the employees through specific fire exists to given protective places, which are not engaged in the fire fighting activities.
4.7.5 Turn off the main electricity from the main panel.
4.7.6 Turn off all the electrical equipment, machinery, and other utensils connected to electricity.
4.7.7 Pool all the fire extinguishers and select which are to be used based on the fire.
4.7.8 Employ the trained firefighters to quickly stop the fire while controlling the spread.
4.7.9 If the fire spread quickly and uncontrollably, call the fire fighting unit to get their support.

4.8 Control of Emergency Power Failures

4.8.1 Once a power failure is experienced, check the main electrical supply and trip-switch for breakdowns.
5.8.2 If the main switch is not tripped, check for the sectional trip-switches for any failures. If there are any failures or breakdowns, inform the Electrician for clarifications and repair of the system immediately.
5.8.3 If the power is disconnected and the backup generator needs to be connected, first check the water level and fuel before starting it. If there are shortages of water and fuel, first refill them up to the standard filling levels.
5.8.4 Follow the given instructions on starting the generator and connecting it to the mainline.

5.8.5 Adjust the circuit breaker to connect the mainline once the generator is operating.

5.8.6 Check the gauges and indicators to understand if the generator is properly functioning.

5.8.7 Power on the main switch.

5.0 References

5.1 Procedure for the handling of potentially unsafe products – FSMS 02/FSP/07

6.0 Records

6.1 Training and testing on emergency situations – FSMS 04/FSP/05/F/01

6.2 Emergency drills – FSMS 04/FSP/05/F/02

6.3 Reports and records on actual emergency situations – FSMS 04/FSP/05/F/03

References

1. British Retail Consortium - https://www.brcglobalstandards.com/
2. British Standards Institute - https://www.bsigroup.com/en-GB/
3. Canadian Food Inspection Agency - http://www.inspection.gc.ca/eng/1297964599443/1297965645317
4. Chambers, A. F., 2007, 'ISO 22000 Food Safety Management Systems: An easy-to-use checklist for small business; Are you ready? Geneva: International Trade Centre.
5. Codex Alimentarius - http://www.fao.org/fao-who-codexalimentarius/home/en/
6. DNV GL - https://www.dnvgl.com/
7. European Food Safety Authority - http://www.efsa.europa.eu/
8. FDA Food Safety Modernization Act - https://www.fda.gov/Food/GuidanceRegulation/FSMA/
9. Food and Agriculture Organization - http://www.fao.org/home/en/
10. Food and Drug Administration of United States - https://www.fda.gov/
11. Food Safety Magazine - https://www.foodsafetymagazine.com/
12. Food Standards Australia New Zealand - http://www.foodstandards.gov.au/Pages/default.aspx
13. Global Food Safety Initiative - http://www.mygfsi.com/
14. Health and Safety Authority of Ireland - http://www.hsa.ie/eng/
15. Health Canada - https://www.canada.ca/en/health-canada.html
    - https://www.iso.org/store.html
16. International Organization for Standardization (ISO) - https://www.iso.org/home.html
17. International Register of Certificated Auditors - https://www.quality.org/
18. International Trade Centre - http://www.intracen.org/
19. ISO 22000 Resource Center - http://iso22000resourcecenter.blogspot.com/
20. ISO 22000:2005, 'Food Safety Management Requirements for Organization in the Food Chain,' first edition.
21. ISO 9001:2015 standard
22. Japanese Standards Association - https://www.jsa.or.jp/en/
23. Mohammed, M. T. Z., Roberts, G, R., Samaraweera D. S. A. and Zoyza, A. K. N., 2008, Hand Book on Tea: Tea Research Institute Sri Lanka.
24. SGS Group - https://www.sgs.com/
25. Sri Lanka Standards Institute - http://www.slsi.lk/index.php?lang=en
26. Sri Lanka Tea Board - http://www.pureceylontea.com/
27. United States Department of Agriculture - https://www.usda.gov/
28. World Health Organization - http://www.who.int/

www.ingramcontent.com/pod-product-compliance
Lightning Source LLC
LaVergne TN
LVHW070149111126
852275LV00002B/2
9789553584007